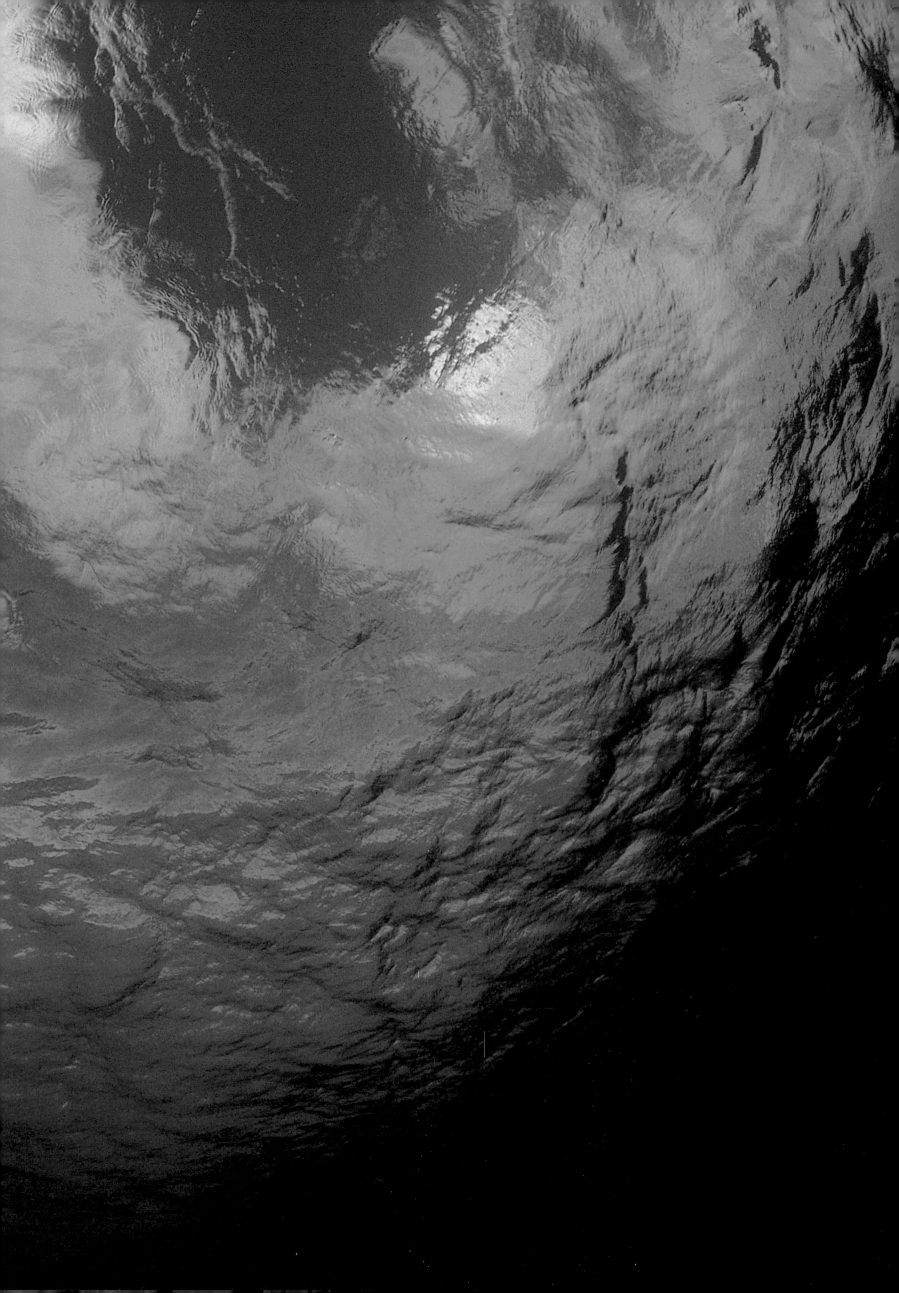

# Deep Sea Odyssey

*Photographs by* **Sophie de Wilde**
*Text by* **Yves Paccalet**

HACHETTE illustrated

# Contents

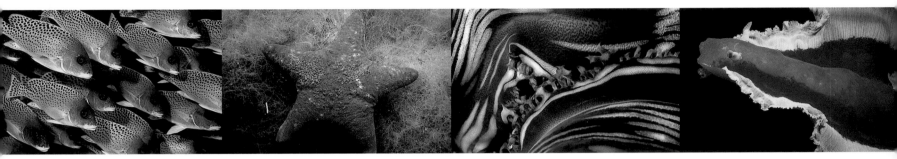

# THE BEAUTY OF LIFE

Photographer Sophie de Wilde

looked beneath the surface of the sea...

and our view of it changed completely.

Sophie: the artist, and the Mermaid... No nickname has ever suited
a woman better. She had the grace of an ocean nymph, the finesse of
a water sprite. To see her swimming below the surface, united with the
liquid element, one might have imagined a creature from the *Odyssey*.
She was like one of the 'amas', the women pearl fishers of Japan.
Sophie de Wilde knew her sea. She lived it body and soul. She knew
it through her eyes and intellect, but above all through her heart...
Among divers she quickly became a figure, a personality, a legend.
She was the best of divers. And the most subtle of photographers.
In communion with the marine element, an eye below the surface...

Sophie fell in love with the sea when she was a child. She learned to
dive holding her breath, then using an aqualung. She studied graphics
and fine art, then worked in advertising and for magazines. One day she
decided to make the sea her profession and her *raison d'être*. Her camera
became her means of communication. She expressed her fascination for
submarine life, her happiness under the sea, on film. She observed,
admired, framed, shot and offered the essence of her surprise and wonder
to the public. The diver embraces the water. She visits the algae, the seashells,
the squid, the fish; the orange sea stars and the violet sea urchins; the
elaborate murex and the gaudy sea-slugs; anemones and spiral tube-worms
with their plumes of tentacles...

'It was in the Maldives, in the Indian Ocean, that my whole existence
changed,' explained Sophie. What fascinates me about the sea is the enigma
of its millions of crossing destinies. The fantasy of evolution, the strangeness
of form and colour, the harmony of everything... I invite myself to the
spectacle and try to share it with others through my photography.
I have passed hours shivering in cold water finally to see in my viewfinder
a little golden seahorse on a bed of blue algae... I have taken what some
call 'risks', diving alone to more than 60 metres (200 feet), to discover the
unreal stripes of an enamoured squid... And as my diving continued, I felt
the need to tighten my field of vision; this was how I started in a field nobody
had taken up: submarine macrophotography. For me, the luminous sea
squirt, the pastel pink shrimp, the minuscule ivory cowrie, the ctenophore as
transparent as Venetian glass, or the nudibranch that looks like a combination
of paintings by Matisse and Picasso!'

So is sketched a destiny. So is a life organised. Sophie offered to others a minute, submerged splendour that she personally knew how to capture. She portrayed the obscure and the lowly, more beautiful than princes and princesses. She was an innovator in this style of picture taking, applying her training in graphics and her woman's viewpoint to her knowledge of the environment. In a domain that, until then, had been the reserve of men, she commanded the respect of the professionals. She was one of the few women in the world – and the only one in Europe – to practise her profession. She has dived more than a thousand times (once a day for three years!), at innumerable sites: the South Atlantic, the Caribbean, the north Atlantic, the Mediterranean, the Red Sea, the Indian Ocean, south-east Asia, the Great Barrier Reef, Polynesia, California...

Sophie de Wilde has become an example. A reference. She published articles and books, held gallery exhibitions, spoke on the radio and on television. She wished to extend her art still further. She wanted to ally macro- with wide-angle photography, to invent a dozen new ways of exalting the splendours of the deep.

She was terrified by the speed at which the marine depths are being sullied, dirtied and destroyed by humans. She had learned to convert her emotion and anger into a militant passion. The protection of nature, and in particular submarine fauna, had become her battle. It was a battle she took to extremes using her own aesthetic weapons.

She was ready to go to the ends of the earth. She wanted to find the wreck of the *Bounty* on the seabed in the Pitcairn Islands. And dive at Easter Island... But destiny brought her an absurd end, near Marseilles. Sophie has escaped into the Big Blue. She has gone to join the jellyfish and the salps, those strange comets whose silver light pulses with the rhythm of the seas and the moon. Her diving friends have placed a commemorative plaque on the seabed at Grand Congloué in the Marseilles Islands, close to the Cassis stone. It is by a drop-off studded with flowers of red coral, renamed the 'Val Sophie' by those who knew of her work and her passion. The inscription there reads 'Sophie de Wilde, marine photographer, 1950–1999. One only truly sees life when seeing its beauty.' Something she was fond of repeating. And which epitomises her philosophy.

Paris, January 2002

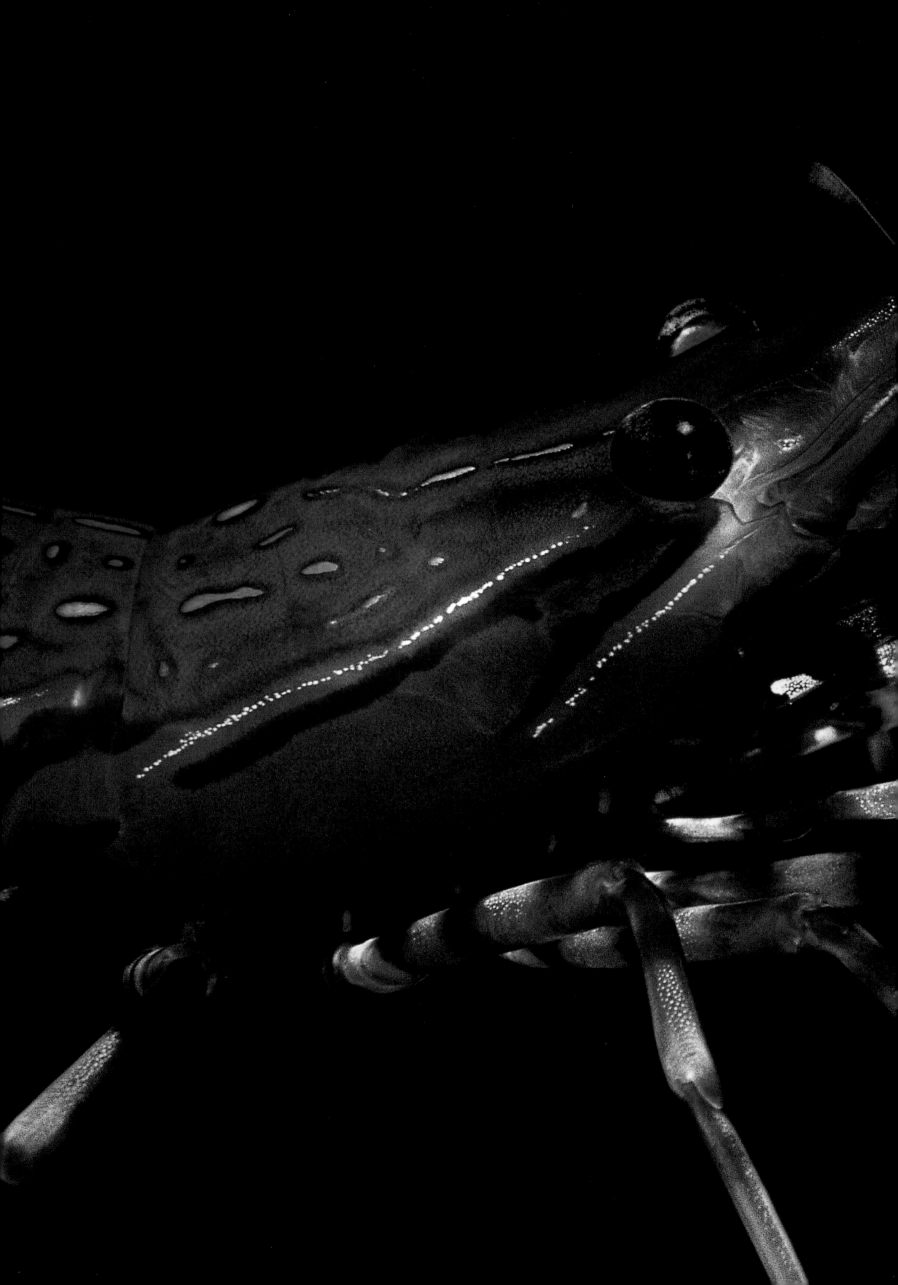

# THE FALKLAND ISLANDS

The Roaring Fifties,

*At Sea Lion Island, in forests*

chilling and mysterious

*of giant algae, the subantarctic seabed harbours*

in the cold depths...

*secret and beautiful flora and fauna.*

# THE FALKLAND ISLANDS

The Roaring Fifties. The anger of the sea... No-one, or almost no-one, has ever dived in these mysterious, icy waters. Such places do not seem created for humans. This is precisely why Sophie wants to dive here!

The Falkland Islands... An archipelago lost in the great South Atlantic, opposite Patagonia, at a latitude of more than 50 degrees south. There are two main islands, East and West Falklands. Strings of rocks whipped by storms, undermined by breakers, frozen by blizzards. The British call them the Falklands, the Argentinians 'Las Malvinas' and the French 'Les Malouines'. Discovered by the Englishman John Davis in 1592, they were revisited by Louis-Antoine de Bougainville in 1764. They are grim, severe, wrapped in mists, rain or snow, ringed by waters that are grey-green rather than blue. To dive here needs courage. The photographer has prepared for her journey with care, aware that she will be among the first to dive here. Aboard a motor vessel hired in the capital, Port Stanley, she adjusts her suit. She warms her hands, wets her mask, checks the gas bottle and regulator. The temperature has fallen during the night. Rain stings the face.

## A FOREST OF GIANT ALGAE

The small island Sophie has chosen for today is called Sea Lion, to the south of East Falkland. In fact, there are sea lions swimming close to the point of embarkation. Our diver has already visited a few sites in the archipelago, somewhat randomly to start with: Volunteer Point, Kidney Island... then organised her exploration, very often being obliged to allow for the difficult weather conditions. The Lively Islands in Choiseul Sound; Barren Island, George Island and Tyssen Island in the Falkland Sound...

She takes her camera and flips over into the sea. The water glows a strange blue-green. The photographer allows herself to be engulfed in the mystery. A fascinating strangeness. She penetrates a world of giant seaweed. A forest beneath, the liquid kingdom of kelp. The life in these cold seas is bizarre. There are few bright colours, no fireworks from angel fish or butterfly fish. A littoral snail with its spiral, pink-striped shell welcomes the visitor. Sophie swims slowly. As she sinks from the vault of this marine temple, one would swear that she is flying, not swimming. The sunken forest of the great South is a magnificent environment in which each being has its exact place. The giant algae are organised in three levels or belts. Down to 3 metres (10 feet), the durvillea resists the breaking waves and the tide. These species are attached to the rocks by means of holdfasts. Their stipes (stems) are short, thick and strong. The undulating fronds (leaves), from 3 to 8 metres (10 to 26 feet) long, divide into dark green straps washed with brown-yellow.

PRECEDING
DOUBLE-PAGE SPREAD
The Patagonian red shrimp (Pleoticus muelleri), its carapace and legs streaked with blue, flourishes its spiny rostrum to daunt its adversaries.

Length: 8 cm (3.1 in)
Depth: 25 m (80 ft)
Site: Volunteer Point

OPPOSITE PAGE
The pink sea-slug (Hermissenda sp.) on a large reddish sponge, confronts possible predators with its poison-charged fingers.

Length: 7 cm (2.8 in)
Depth 30 m (100 ft)
Site: Kidney Island

ABOVE, LEFT
A colony of tiny, still-unidentified sea squirts of the clavelinid family.

Length: 2 cm (0.8 in)
Depth: 35 m (115 ft)
Site: Sea Lion Island

ABOVE, RIGHT
A bivalve mollusc Gaimardia trapesina patrols the immense oarweed.

Length: 3 cm (1.2 in)
Depth: 25 m (80 ft)
Site: Sea Lion Island

*The sunken forest,
a splendour where
each being occupies
an exact place.*

Uprooted, they drift off like improbable sea-snakes.

Between 3 and 6 metres (10 and 20 feet) down, where the forest is less dense, the lessonias take charge. Here there are some other plant species: the reddish blades and garlands of delesseria; the delicate green sails of sea lettuce; the tousled mane of desmaretia, brown-red shot with gold.

### THE SPIRITS OF THE MAGIC FOREST

The macrocyst level extends down from 6 to 25 metres (20 to 80 feet). These brown, laminar algae of the lessonia family are known to science

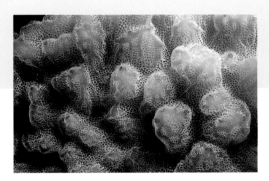

as 'pear-bearing macrocysts' (*Macrocystis pyrifera*). They are the colossi of the plant kingdom and may be as long as 50 metres (165 feet). Despite their gigantic size, they are annual plants: they are capable of growing by 50 centimetres (20 inches) per day! The thin, elegant stipe rises in an arc, then flattens and divides above; each blade in turn divides into toothed leaflets whose bases are swollen with vesicles. These swellings suggest the imminent birth of some still unknown animal. Filled with air and carbon dioxide, they serve for flotation.

Sophie strokes a frond. The strange softness of kelp, a touch that reminds one of a baby's skin... She pushes in between the stems, becoming a sea lion, sinuous as a cormorant. One is reminded of Beaudelaire's Correspondances 'Nature is a temple with living pillars...' The dull blue-

green water is forgotten. Closer to the bottom, the water darkens. The diver slips between the submerged 'trunks'.

In this silent forest, the power of the macrocysts has the presence of oaks and beeches on dry land. A series of illusive images created with variegated greens, blues and greys. Here, a medieval storyteller might invent legends.

Sophie brushes against a rock carpeted with red algae — *Lithothamnes* and *Lithophyllus*, in ridged layers. A hermit crab camouflages its borrowed snail shell with pieces of mussel and weed: each seems disguised — the crustacean as a rock and the rock as an arthropod. *Loxechinus* and *Abacia* sea urchins await nightfall before climbing to browse the fronds. Each giant alga is spangled with a coat formed by thousands of other algae, many of which will not reach more than a centimetre (0.4 inch) in length. They are not parasites, but epiphytes: these dwarves use the kelp as a kind of perch that brings them closer to the light. They serve as pasture for armadas of invertebrates: waving worms, clicking crustaceans, shining sea urchins... and the seedlings of tiny orange sea cucumbers or holothurians, the largest of which is no more than 2 centimetres (0.8 inch) long. And hordes of the yellow bivalve (*Gaimardia trapesina*), in places so thick that even the giant kelp bends under their weight.

### A NOVEL DISPLAY

Between 20 and 25 metres (65 and 80 feet), the macrocysts are not the only plants to act as submarine trees. Mingled with them are giant lessonias over 30 metres (100 feet) long, their 'trunks' some 15 centimetres (6 inches) in diameter. Sophie reaches the bottom. The rocks and kelp holdfasts form a labyrinth of hiding-places, fissures, corridors and dormitories hosting numerous

PRECEDING
DOUBLE-PAGE SPREAD
The Falklands hermit crab *(Pagurus sp.)* hides its delicate body in a carefully decorated, camouflaged snail shell.

Length: 10 cm (4 in)
Depth: 15 m (50 ft)
Site: Salvador

ABOVE
This as yet unidentified colony of sea squirts may belong to the Stolonica family.

Length 2 cm (0.8 in)
Depth: 25 m (80 ft)
Site: Kidney Island

OPPOSITE PAGE
An orange sea cucumber *(Cucumaria sp.)* spreads the tree of tentacles directing plankton and organic debris into its mouth.

Length: 2 cm (0.8 in)
Depth: 40 m (130 ft)
Site: Sea Lion Island

creatures. The tubes of annelid worms (peacock and spiral tube-worms) shoot fireworks of feathery tentacles. Bryozoans take on the semblance of a goldsmith's niello. Bivalves yawn — great blue mussels, rounded cockles... climbing gastropods: abalone, snails, sea-slugs, and murex with its barbed spike. A profusion, a superabundance of beings... The variety of forms, tints, behaviours, regimes, methods of reproduction seems infinite... Of the crustaceans, the squat- and krill-lobsters are the most abundant. The former, cousins of the hermit crab, have no need of a borrowed shell for protection: they swarm clad in orange armour set off by transverse blue waves... The krill-lobsters with their long legs and powerful claws look like shrimps, and are a main food item for the sea lions.

Sophie wanders through the submarine forest. Elegant, shapely, supple, a perfect swimmer, like the sea lion that now dives to brush past her, curious, almost friendly. Complicity between mammals: warm blood in a cold sea.

### SOME STRANGE FISHES

A strange dive... Nothing looks like anything else; but all is magnificence... Some squid ripple their unreal colours like neon advertising. Great sea-stars — suns with 20 branches — climb about searching for bivalves to open and digest on the spot. Resting on the rocks are other stars of the genera *Porania* and *Cyreihra* with their radiating symmetry and the enigma of their existence. Among the most common animals are the brittle-stars; these snake-like sea-stars writhe about displaying their spindly arms. Hundreds of these creatures gather together, like Medusa with her head of serpents. Just as one meets with few large animals in the woods on land, so are fish rare

amongst the kelp. Most are nocturnal. In the daylight hours they hide or rely on mimicry. Sophie gently extracts a small *Harpagifer georgianus* from a bunch of seaweed 'root'. Eight centimetres (3 inches) long, it is beige with red streamers provided with a spine on each gill. The creature does not attempt to escape,

allowing itself to be returned to the protection of its cave by the current. It embodies simplicity and innocence, a witness to a time not long gone when humans were unknown here. Some black rockfish pass. A zoarcid (eelpout), the curious lizardfish (*Lycodichthys antarcticus*) with its reptilian head and long, brown-yellow striped body, accepts the diver's teasing. Sophie is still looking for a hagfish, that strange, primitive, jawless fish (of the group Agnathes) cousin to the lamprey, which sucks up its food through a funnel-shaped mouth. The best-represented family in these parts is that of the notothenids, or Antarctic morays, though these creatures are far removed from the true moray: they are closer to the weever fish. With serpent's head, globular eyes, prominent lips and spiny back, these fish are no paragons of beauty! But they carry with them the mystery of the improbable.

The sea lion returns to twirl beneath the photographer. Collusion in contact. Harmony between two bodies. Unseen and in unison, the two sirens dance in the cold of this sea at the end of the earth.

**PRECEDING DOUBLE-PAGE SPREAD, LEFT**
The fronds of kelp *(Macrocystis pyrifera)* float and ripple in the current, buoyed up by egg-shaped bladders filled with carbon-dioxide-rich air.

Length of a follicle: 50 cm (20 in)
Depth: 5 m (16 ft)
Site: Kidney Island

**PRECEDING DOUBLE-PAGE SPREAD, RIGHT**
The primitive sea-slug *(Tylodina sp.)* wanders over the kelp with its miniature shell, bulky foot and abnormally large tentacles.

Length: 5 cm (2 in)
Depth: 15 m (50 ft)
Site: Volunteer Point

**OPPOSITE PAGE, ABOVE**
The pycnogonid decorator sea-spider *(Achelia sp.)* perches on an alga. This primitive arthropod, a cousin to the spider, is a survivor from the Palaeozoic era.

Length: 2 cm (0.8 in)
Depth: 25 m (80 ft)
Site: Sea Lion Island

**OPPOSITE PAGE, BELOW**
The white ostrich-plumed hydroid *(Aglaophenia sp.)* presents a bristling lacework of miniature polyps clad with stinging tentacles that paralyse the plankton.

Colony height: 5 cm (2 in)
Depth: 20 m (65 ft)
Site: Kidney Island

**ABOVE**
The bivalve mollusc *Gaimardia trapesina* feeds on animalcules it finds on the kelp.

Length: 3 cm (1.2 in)
Depth: 25 m (80 ft)
Site: Sea Lion Island

# THE BAHAMAS

The light,

*Angels mingling with devils:*

the perfect light

*a bizarre fauna amongst*

of the Caribbean Sea!

*the deer horn coral and sea fans...*

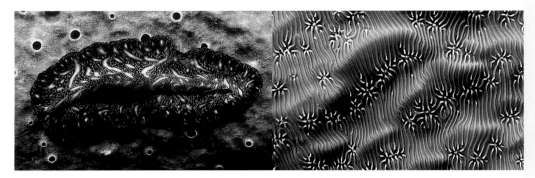

Pitiless splendour… the frigate birds wheel in the clean sky. The brown gannets, the red-footed gannets, the brown pelicans dive on a school of sardines tracked from below by fork-tailed jacks and gold and jade dolphin fish. A swirl: a whale shark opens its maw. A manta ray leaps: the two-horned 'sea-devil' returns to the water's graceful embrace. The little boat rocks. Coral beneath the hull, coconut palms in the distance, the soft trade wind: no remote connection with the cold of the Falklands. Our diver has traversed the Bahamas archipelago (Andros, Long Island, Bimini Island)… She has established herself in the Turks and Caicos Islands. With her photographer's eye she has scanned Belize (the immense Chinchorro Bank), Honduras and Guadeloupe, and now wants to explore the Turk Island Passage between Grand Turk and the Mouchoir Bank (an extension of the Silver Bank), where many galleons of the Spanish treasure fleet slumber. According to marine ecologists, these areas belong to the Caribbean domain which, strictly speaking, includes the Gulf of Mexico and the western Atlantic from Florida to the north of Brazil. Here can be savoured the mildness of what was once called the West Indies.

## FOLLOWING THE ANGEL FISH

A French angel fish — (*Pomacanthus paru*) emerges from the coral: on its coal-black flanks each scale shines, a silver crescent of light. The animal surfaces and exposes its body for a moment, then dives back. It would be impossible not to see this gesture as an invitation. Sophie dives over the side. She swims toward a spotless seabed of white sand between two walls of coral. This place might deserve the name 'Happy Valley'. There is a winding corridor carpeted with a sediment of sparkling white. The main corals that line the passage are deer horn (*Acropora cervicornis*) and moose horn (*Acropora palmata*), so called because of their branched polyparies, their extremities pointed in the former and flattened in the latter. The photographer passes along this rampart of branched coral. It suggests Atlantis: in the Bahamas, the mysterious submarine walls of the Bimini Isles have been likened — by lovers of the fantastic — to the three walls (bronze, brass and gold) of that submerged city. Many of the reef-building corals are branched or cushion-shaped, like blue coral (or porite); pencil coral, and the spiralling coral that forms Sleeping Beauty castles. Some, the star coral *Monastrea cavernosa* for example, form crusts; others are mushroom-shaped. The bizarrely formed brain coral (*Colpophyllia natans*) might lead you to believe that the sea can think. Here are fleshy corals, leafy corals, artichoke coral and all sorts of coral flowers (tube, scarlet and gold star) with calyces of gold or of blood, orange or amethyst. Some of the products of evolution hint at a kind of madness.

PRECEDING DOUBLE-PAGE SPREAD
The lettuce nudibranch (*Tridachia crespata*) undulates over a red sponge, searching for the caulerpa algae on which it grazes.

Length: 4 cm (1.6 in)
Depth: 10 m (33 ft)
Site: Grand Turk

ABOVE, LEFT
A tropical marine worm — a coloured flatworm, probably of the genus *Pseudoceros*.

Length: 3 cm (1.2 in)
Depth: 25 m (80 ft)
Site: Andros

ABOVE, RIGHT
Part of the polypary of a leaf coral, showing the star-shaped polyps.

Length: 5 cm (2 in)
Depth: 45 m (150 ft)
Site: Long Island

OPPOSITE PAGE
Fronds of the brown sargassum weed (*Sargassum* sp.) showing the floats that allow it to drift in the surface current.

Length: 20 cm (8 in)
Depth: 1 m (3.3 ft)
Site: Andros

Beside the reef-building polyps or madrepores (true corals, otherwise termed 'scleractinia', 'hermatypic', 'hard' or 'stone' corals: in the poetry of science!) the diver discovers fire corals — the millipores, against which it is best not to brush the naked skin: the poisonous cells of these species can inflict cruel burns. As might a queen angel fish or a four-eyed butterfly fish, Sophie wriggles between the soft corals (Alcyonaria) resembling orange, pink or purple Christmas trees. She passes banks of ochre, red or lilac sea fans, and glides over twists of pallid sea whip gorgonias with names like 'wire' or 'white-haired' that fit them well.

### A HUNDRED BEAUTIES REVEALED

The photographer zigzags through a maze of large Venus' fan gorgonias, with branches alternating in colour between shades of violet, pink and beige. The lacework formed by these creatures is typical of the Caribbean. Sometimes as much as a metre (more than 3 feet) high, it provides a haven or hunting ground for thousands of other beings. With her eye to the viewfinder and her lens aimed at these sublime baubles that almost nobody has ever seen, Sophie frames a stenorhynic ('narrow-nosed') arrow crab with spindly legs; a decorator crab that has covered itself in algal debris and shells; a transparent cleaner shrimp with mauve markings; and a little snail (*Cyphoma gibbosum*), a sort of oval the size of a finger-joint, also known as the 'flamingo tongue snail', whose cream-coloured shell is set off by violet eye-spots with ochre centres.

Each detail is like a picture by Chardin. For example, the sponges… These are the simplest, most primitive multicellular animals. Ochre, violet or yellow tube sponges (or organ pipes) *Aplysina* and *Agelas* vie with rose- or mauve-coloured vase sponges. Bowl, ball, encrusting and branching sponges give way to giant barrel sponges (*Xestospongia muta*), sometimes as high as 1.80 metres (6 feet), into whose central cavity the diver glides with delight.

Fixed in the ground and highlighted with silver, hydroids play their part as fine-barbed sea-pens. Bushes of black corals (Antipatharia) sway about in the darker parts of the reef. Jellyfish pulse in the open water. Sophie avoids the tentacles below the mauve float of the physalis (Portuguese man-of-war). Deadly poison!

Everywhere invertebrates gape, fidget, ripple, climb, shoot along, pace up and down, jet about, each according to its particular method of locomotion. The giant anemone (*Condylactis gigantea*) reaches 30 centimetres (12 inches) in diameter

and imitates a Medusa's head with green and mauve hair. *Pseudoceros* flatworms pass, their cut and colours as taken from a medieval illumination. On the rocks, giant Christmas tree worms (*Spirobranchus giganteus*) shoot out feathery yellow, blue or red fireworks.

### VAGRANT LOBSTER

Sophie descends to a coral corner where the variety of life is exploding. A red and white banded coral shrimp (*Stenopus hispidus*) is on parade with its golden cousin. A service station is open: a cleaner shrimp removes parasites from the mouth of a striped grouper. As its name indicates, the anemone shrimp squats on the sea anemone (actinia). The shimmering transparency of a

*The photographer lingers over a hundred beauties discovered through her curiosity. Each detail is like a painting by Chardin.*

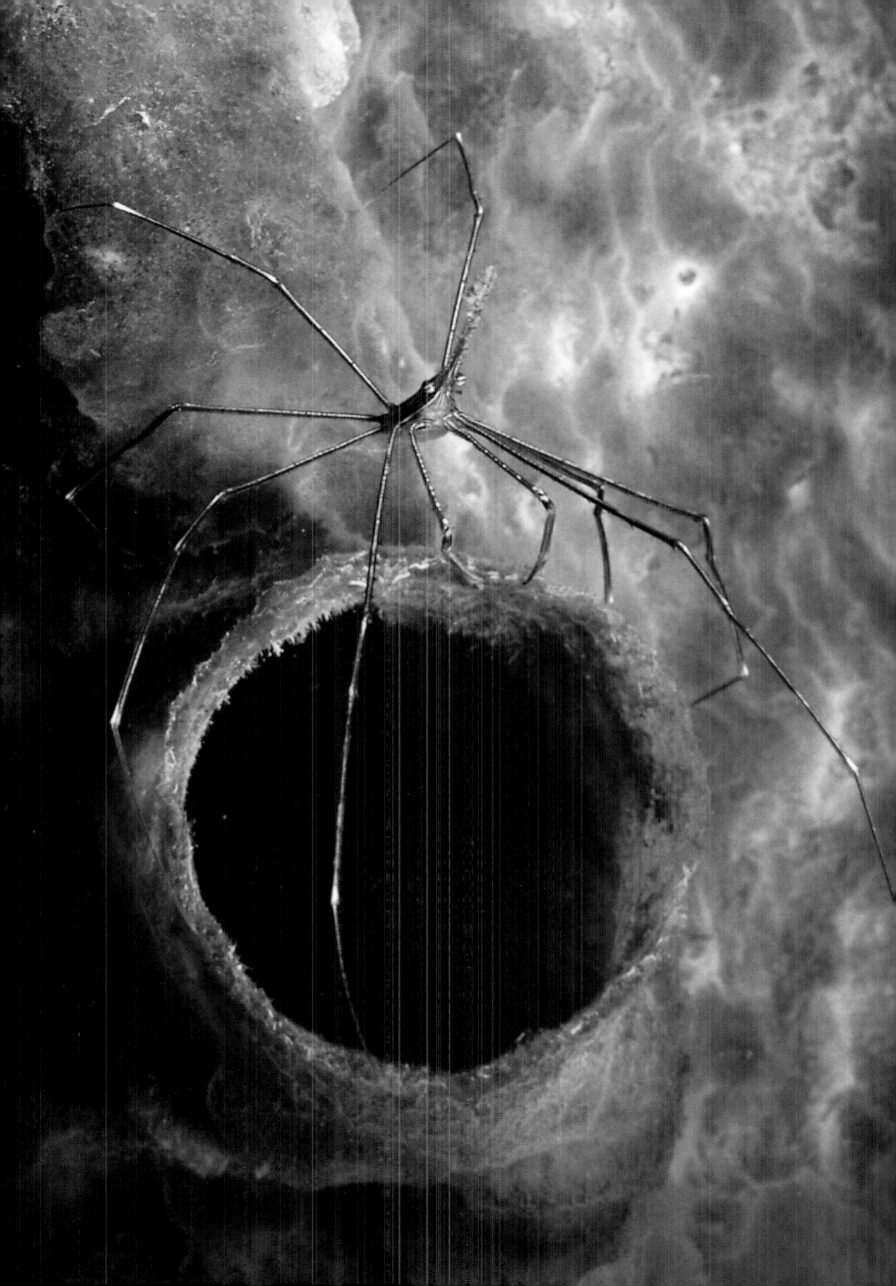

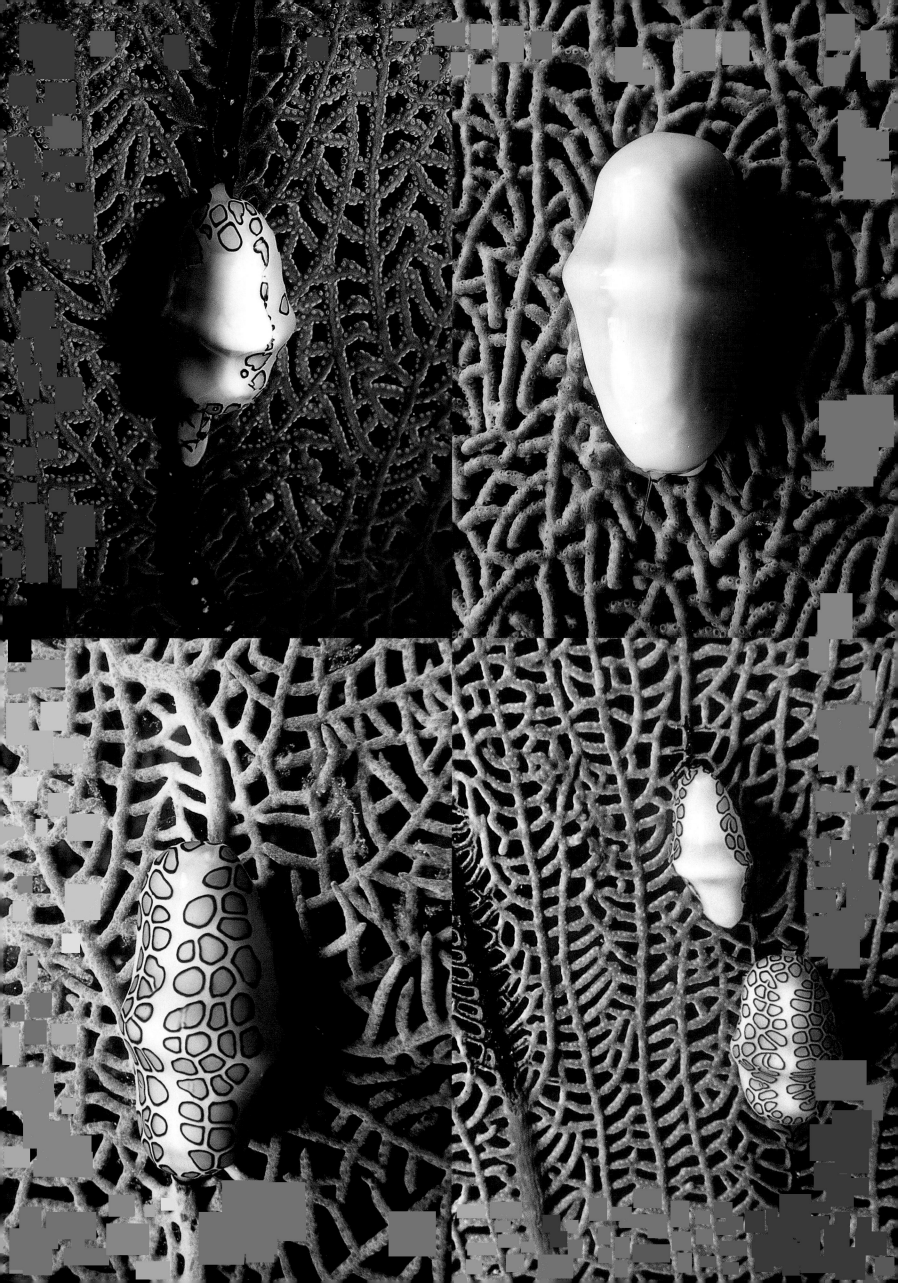

Yucatan anemone shrimp elicits admiration. In a hole, the Caribbean spiny lobster (*Parulirus argus*) whips the water with its antennae. In winter it will migrate with its fellows in Indian file to the dark mystery of the deep seabed. Here and there other crustaceans reveal themselves: mantis shrimp (squill) with their flattened, shovel-like antennae, hermit crabs in their salvaged shells, cowrie crabs, bearded clown crabs, box crabs, decorator or ghost crabs. And mantis shrimp with raptor arms, hunting the coral just as their religious namesakes hunt the bushes on land.

Sophie photographs the precious lace of five different bryozoan colonies in succession. She identifies some molluscs. The king of sea snails is the queen conch (*Strombus gigas*), sometimes as long as 50 centimetres (20 inches). Tulips and tritons announce olives and ceriths, venomous cone shells, cowries coloured more delicately than the most beautiful objects made by human craftsmen… Among the gastropods are the sea-slugs — the nudibranchs (*Chromodoris, Elysia, Facelina,* etc.), as gaily coloured as a convention of harlequins. Alongside some lamellibranchs (bivalves) gape flame scallops, close relations of the St Jacques or great scallop, their mantles bordered with tentacles and rows of blue eyes. Considering such animals, one might say that the sea is looking at the diver… Looking at the cephalopods, can we ignore the roundabout of squid, two-spotted bumblebee octopus (*Octopus filosus*) and their cousins, the common reef octopus, the pygmy, and that one with white spots? The sea-lilies (crinoids) represent the echinoderm (spiny skin) family. Here too are holothurians or sea cucumbers; ophiurida or brittle-stars; sea urchins (pencil, diadem, purple heart, sand dollar…); and starfish; the thick star cushion, the one with nine arms, the astropecten or sand star, the blue,

comet-shaped *Linckia*… Tunicates, such as the multicoloured ascidians (clavelinids, etc.), lead along the path of evolution to the vertebrates.

## SHARKS AND SURGEON FISH

The fish win! Sharks (reef, white-tipped, nurse) and rays (eagle, torpedo, stingray…). Barracuda, amberjack, tarpon and crevalle jack. Morays, seahorses, grunts, goatfish, trigger fish, tropical sole, poisonous scorpion fish. Porcupine fish, their bodies clad in needles. Trumpet fish with Cyrano noses. Angler fish with 'Hot Lips' mouths. Batfish. Rainbow wrasse, soldier fish and squirrel fish… And coral-eating parrot fish, their great teeth fused together. The grouper

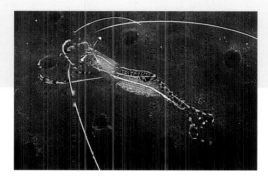

(striped Nassau, diamond) and dragonfly fish. Red lutjans, graceful surgeon fish, angel fish and butterfly fish. A shimmering of scales. A language of colours, lines, motifs, friezes, stripes and spots, ever incomprehensible to humans. Farther on, almost indistinct in the blue, a marine turtle rows majestically on its way. A thought for the Caribbean monk seal, wiped out by humans, now no more than a blurred image in the blue. A group of spotted spinner dolphins brings an end to the dive: mobile aquatic splendour. Live torpedoes graceful speed in the stream of life… The cetaceans surface at the same time as the photographer. With her, they offer the world the universal symbol of breath. Like the breathing of a poet born to the sea…

*Living torpedoes, graceful speed in the stream of life…*

PRECEDING DOUBLE-PAGE SPREAD, LEFT
The pink and beige majid decorator spider crab *(Microphrys bicornuta)* fulfils his simple role, in ambush on a vase sponge *(Spinosella plicifera)*.

Length: 2 cm (0.8 in)
Depth: 40 m (130 ft)
Site: Roatan Island

PRECEDING DOUBLE-PAGE SPREAD, RIGHT
Unperturbed by the diver, a minuscule arrow crab *(Stenorhynchus seticornis)* waits in ambush on a large vase sponge *(Spinosella plicifera)*.

Length: 2 cm (0.8 in)
Depth: 35 m (115 ft)
Site: Roatan Island

OPPOSITE PAGE
The delicate flamingo tongue snail — also known as the Caribbean money cowrie — climbs on a mauve common sea fan *(Gorgonia ventalina)* where it flushes out its food. Subtle colour variations…

Length: 1 cm (0.4 in)
Depth: 15 m (50 ft)
Site: Grand Turk

ABOVE
This Pederson cleaner shrimp *(Periclimenes pedersoni)* has chosen a red crust sponge for its home and hunting ground.

Length: 2 cm (0.8 in)
Depth: 10 m (33 ft)
Site: West Caicos

# IRELAND

*At Connemara, the lobster in blue Celtic armour,*

## The long, silver swell

*the spiny northern starfish*

## and warm caress of

*and the shadowy moray eel pass*

## the Gulf Stream.

*among the brown algae.*

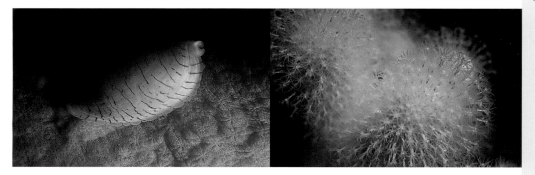

The long swell raised by the wind breaks up into white foam and exhausts itself on the Connemara shore. The magic of grey-green water beneath the clouds. The Celtic greenness of the hills, where the grass and the bracken are studded with pink-mauve heather and yellow, butterfly-flowered gorse. The breakers beat into Clew Bay. On the horizon are outlined the islands of Achill, Clare, Caher, Inishturk, Inishbofin and Inishark. The canoe makes for Caher. Almost a nothing in the sea: some stones, moorland and birds: puffins, small penguins, guillemots, cormorants, seagulls, herring gulls, petrel fulmars and gannets... Seals lie on the rocks, port seals (sea calves) with their sharp noses, those of the grey seals endearingly large. Here everything depends on a colossal stream from the New World. Born in the Caribbean Sea and the Gulf of Mexico, the Gulf Stream travels more than 5,000 kilometres (3,000 miles) to give Europe a warm kiss: 80 million cubic metres (105 million cubic yards) of water per second, 500 times more than the Amazon, a hundred times more than all the rivers on earth! The Gulf Stream softens the climate of the western aspect of the Old World. Thanks to this, Irish and British winters are milder than those of Labrador. Caher Island seems to belong essentially to the ocean. A fisherman lifts his lobster pots. A trawler hauls its net. But there are no diving boats: amateur divers rarely venture here,

in these bays as distant from the equator as the Falklands. The Gulf Stream warms Ireland, but it does not move it to below the tropics! Sophie swims towards some shallows where the waves strike, the wind carrying the spray up onto the coast. The pale eddies spiral like a DNA molecule: together the scientist and the poet decipher the beginnings of life. Each dive is like a profane mass, calling for an introit. In the tidal ('mediolittoral') zone and below ('infralittoral'), the prairie of the ocean is the square before the temple. Some algae are green: sea lettuce in dancers' veils; udotea the ragged mermaid's fan, enteromorpha with naiad hair... Among the red algae is carrageen (*Chondrus crispus*), so useful in medicine; purple callophyllus, bloody delesseria, the red-brown ribbons of vermilion nitophyllum... To these are added the brown algae: pretty fractally branching antler weed, with blue-green fluorescent stripes; the fucus (varech, bladder and spiral wrack) held up by their ovoid floats; the fan-shaped peacock's tail, the greenish desmaretia; the delicately branched bush of cystoceira.

## THE LITTLE HORSE OF THE ALGAE

This jungle conceals timid little fish such as blennies (common, trumpet and spotted) and gobies (black and rock). A syngnathid pipefish (*Syngnathus agus*) undulates: this

PRECEDING DOUBLE-PAGE SPREAD
The common octopus (*Octopus vulgaris*) gracefully spreads its tentacle 'flower'.

Length: 50 cm (20 in)
Depth: 20 m (65 ft)
Site: The Skellings

ABOVE, LEFT
The Arctic cowrie or coffee bean (*Trivia arctica*) feeds on red sea squirts.

Length: 1.5 cm (0.6 in)
Depth: 45 m (150 ft)
Site: Inishbofin

ABOVE, RIGHT
Tufts of the soft alcyonarian coral (*Alcyonium digitatum*), sometimes called 'dead men's fingers'.

Length: 5 cm (2 in)
Depth 25 m (80 ft)
Site: Inis Degil

OPPOSITE PAGE
The tiger flatworm (*Prostheceraeus vittatus*) with its leaf-like, violet-striped body, glides over a sponge using thousands of tiny ventral cilia.

Length: 3 cm (1.2 in)
Depth: 20 m (65 ft)
Site: Carrygaddy

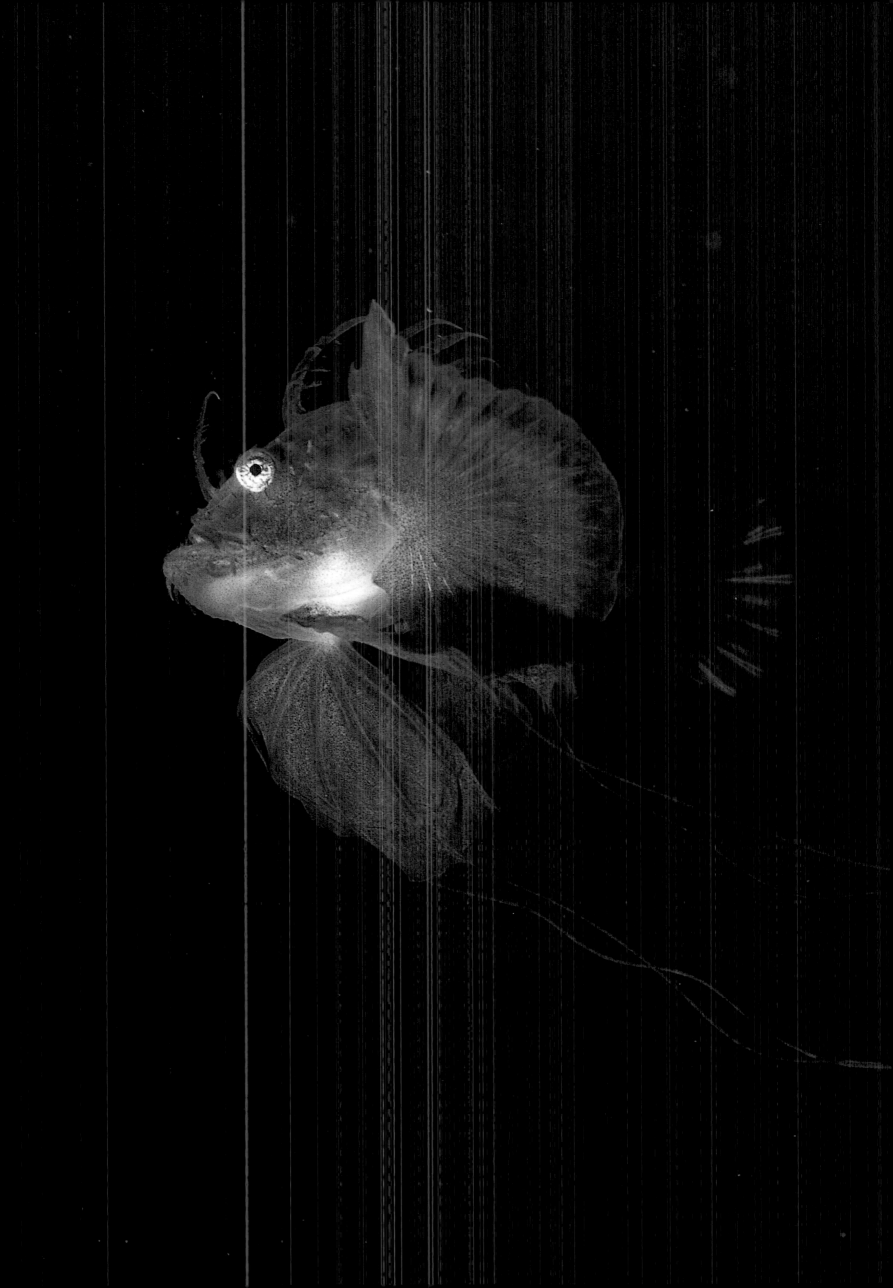

animal, taken by some as a snake and called the 'sea needle', has the body of a reptile; but its fins and horse's head show it to be a fish, a cousin of the seahorse... Just so a seahorse, and here is one, its tail curled round an alga. Its plumed back and starry scales identify it as the long-nosed species (*Hippocampus ramulosus*). Some shrimps jet backwards: grey (*Crangon vulgaris*), northern shrimp, and shrimp... And pink ones, nicknamed 'bouquets' or 'scampi' (*Leander serratus*). Sophie entices a pistol shrimp (*Alpheus ruber*) hiding under a stone: this 3-centimetre (1.2-inch) long crustacean has two stout claws which it snaps in the water, causing the detonations (by cavitation) that give it its name. A double advantage: it frightens predators and stuns small prey.

admirable ecosystem. The photographer lets herself drift into the fantasy of an eddy. Like a sea sprite she penetrates the forest of algae. The North Atlantic kelp is not as huge as that of the southern ocean. But it displays the same forms and plays the same roles. It serves as food for herbivores. It oxygenates the water by photosynthesis. It provides an infinity of hiding places for species, reproducers, eggs and larvae. Between 0 and 40 metres (0 and 130 feet) down, it performs the triple function of food store, lung and nursery. Sophie crosses the shadow of a seal and some other fleeting figures: some cod... These fish, once believed by the fishermen to be more numerous than the pebbles on the beach, have become rare: man the slaughterer is the culprit.

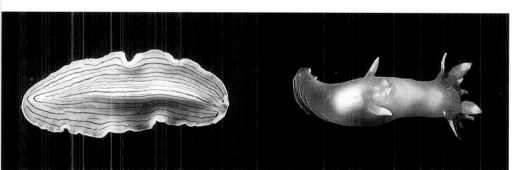

PRECEDING
DOUBLE-PAGE SPREAD,
LEFT
Pulsing in open water, the lion's mane jellyfish (*Cyanea capillata*) captures plankton with its superb mane of highly venomous tentacles.

Length: 50 cm (20 in)
Depth: 2 m (6.5 ft)
Site: Inishbofin

PRECEDING
DOUBLE-PAGE SPREAD,
RIGHT
This baby angler fish (*Lophius piscatorius*) already possesses the worm-like thread with which it will lure its prey.

Length: 10 cm (4 in)
Depth: 45 m (150 ft)
Site: Clare Island

OPPOSITE PAGE
This tentacled coelenterate hanging onto the stipe of a brown laminarian alga (*Laminaria saccharina*) is the 'polyp' or static form of a future wandering moon jellyfish (*Aurelia aurita*).

Length: 2.5 cm (1 in)
Depth: 15 m (50 ft)
Site: The Skelligs

ABOVE, LEFT
The striped tiger flatworm, a turbellarian (*Prostheceraeus vittatus*) climbs on a red alga.

Length 3 cm (1.2 in)
Depth: 20 m (65 ft)
Site: Inishark

ABOVE, RIGHT
The four-lined nudibranch mollusc (*Polycera quadrilineata*) displays its varicoloured design.

Length 3 cm (1.2 in)
Depth: 35 m (115 ft)
Site: Inishark

The diver executes an arc and sinks into the bottle green water shot with the unreal brown-yellow light reflected by the kelp. This world harbours legends. One thinks of St Brendan's voyage in a cockleshell (in the sixth century, it is said) to preach to the Celts. He found an island, landed and lit a fire. The island awoke in a rage: it was a whale; the Leviathan of the Bible. St Brendan begged God to pity him and was saved.

## THE FANTASY OF AN EDDY

The water is loaded with particles. Therein sand, microscopic algae and animalcules flow in a fertile commotion. The ensemble of liquid, mineral and plankton creates an

With its characteristic three dorsal fins, the true cod (*Gadus morhua*) and its cousins the saithe, pollack, capelin and whiting, remind our improvident, plundering species of the ocean's fragility. Under a rock moves an enormous head and the beginning of a snake-like body: a powerful conger in ambush... A blue-clad lobster snaps his claws. A large turtle crab with its ochre and green shell replies. The double crack of a whip: the antennae of a crayfish. Sophie grasps a leaf (frond) of a great brown alga, strokes its stem (stipe) and sinks down to its root (holdfast). Different types of laminaria populate the seabed. Oarweed (*Laminaria digitata*), a metre (3.3 feet) in length, has a stout stipe terminating in a 'hand' with flat 'fingers'. The sugar kelp (*L. saccharina*), 3 metres (10 feet)

long, consists of a single dented, gathered ribbon with sinuous edges. The two most attractive species are tangle (*L. hyperborea*) – up to 3.5 metres (11.5 feet) long, and *Sacchorhiza polyschides*, which reaches 5 metres (16 feet). These colossal plants intermingle their fronds, which are divided into thongs and perched on a thick stipe, the former smooth, the latter embossed. The diver looks over a rock resembling a whale's head, colonised by legions of invertebrates, among which are grey, sac-like Sycon sponges, *Axinella verrucosa* in vermilion bushes, and others in cream-coloured tubes, pink cushions and orange balloons. The anemones wave their tentacles in the current: carmine in the horse anemone, pale green with mauve tips in the common

pale, tree-like soft corals, or alcyoniarians (*Alcyonium digitatum, A. palmatum*), whose polyp-decorated 'fingers' troubled the fishermen: they called them 'mermaids' hands', 'drowned men's hands' or 'dead men's fingers'. There is an appointment with beauty in the purple lines of the pink flatworm (*Prostheceraeus vittatus*). And in the rainbow colours of the swimming nereid worms; or the sedentary annelid worms: the peacock worm, tube-worm and spirograph... The shellfish are splendid too – bivalves and gastropods. On one side are blue mussels, oysters, cockles, Venus shells and clams. On the other are abalone (haliotis, or sea ear), limpets, topshells, periwinkles, whelks... not forgetting the cephalopod molluscs – squid, cuttlefish and octopus.

*The spiny starfish, like a polar star in the firmament of the great deep...*

anemone. The tentacles of the plumose anemone (*Metridium senile*) are like white hair. Others, such as the solid cerianths, sink their long feet into the sediment. Or take a ride on a hermit crab (for mutual protection) – as with the anthozoan (*Adamsia palliata*), which is carried and fed by the hermit crab (*Eupagurus prideauxi*).

## AS FREE AS THE MEDUSAS

Jellyfish pulse in the open water: the lion's mane (*Cyanea lamarkii*) with its 32-lobed umbrella; the chrysaora with glints of jade and gold; the pink and mauve mushroom of the pelagic noctiluca; or aurelia's four violet horseshoe-shaped reproductive organs, clearly visible through its transparent body. On a slope grow

The crustaceans either walk or swim: crayfish, galatea, sea-spiders, box crabs, hairy crabs, ghost crabs and green crabs... The echinoderms pace the seabed on their hundreds of walking legs: sea urchins, sea cucumbers, brittle-stars and starfish (common red starfish, orange sand star, sun starfish with eight to ten arms, etc.). Among these last, the mystery of the North Atlantic recurs. The word 'north' in the 'north star' (*Marthasterias glacialis*) is not there by chance: it has five spiny, grey-green arms... a devourer, hated by those who farm mussels and oysters, it can force open bivalve shells and, extruding its stomach, digest them on the spot. Nonetheless, it is as beautiful as a polar star in the great firmament of the deep!

PRECEDING DOUBLE-PAGE SPREAD
This striped jellyfish (*Chrysaora hyoscella*) balancing its violet-striped umbrella, is also called 'gold tail'. It glows as though lit from within.

Length: 60 cm (24 in)
Depth: 2 m (6.5 ft)
Site: The Skellings

ABOVE, LEFT
The Atlantic lobster (*Homarus gammarus*) in its violet-blue armour seems to be on parade before a tournament.

Length: 30 cm (12 in)
Depth: 15 m (50 ft)
Site: The Skellings

ABOVE, RIGHT
The central mouth and tentacles of the dahlia sea anemone (*Urticina = Taelia felina*).

Length: 8 cm (3 in)
Depth: 45 m (150 ft)
Site: Inishbofin

OPPOSITE PAGE
The strange plumose anemone (*Meridium senile*), also called 'old man's hair' or 'carnation anemone', displays a thousand pale tentacles.

Length: 15 cm (6 in)
Depth: 50 m (165 ft)
Site: Inishturk

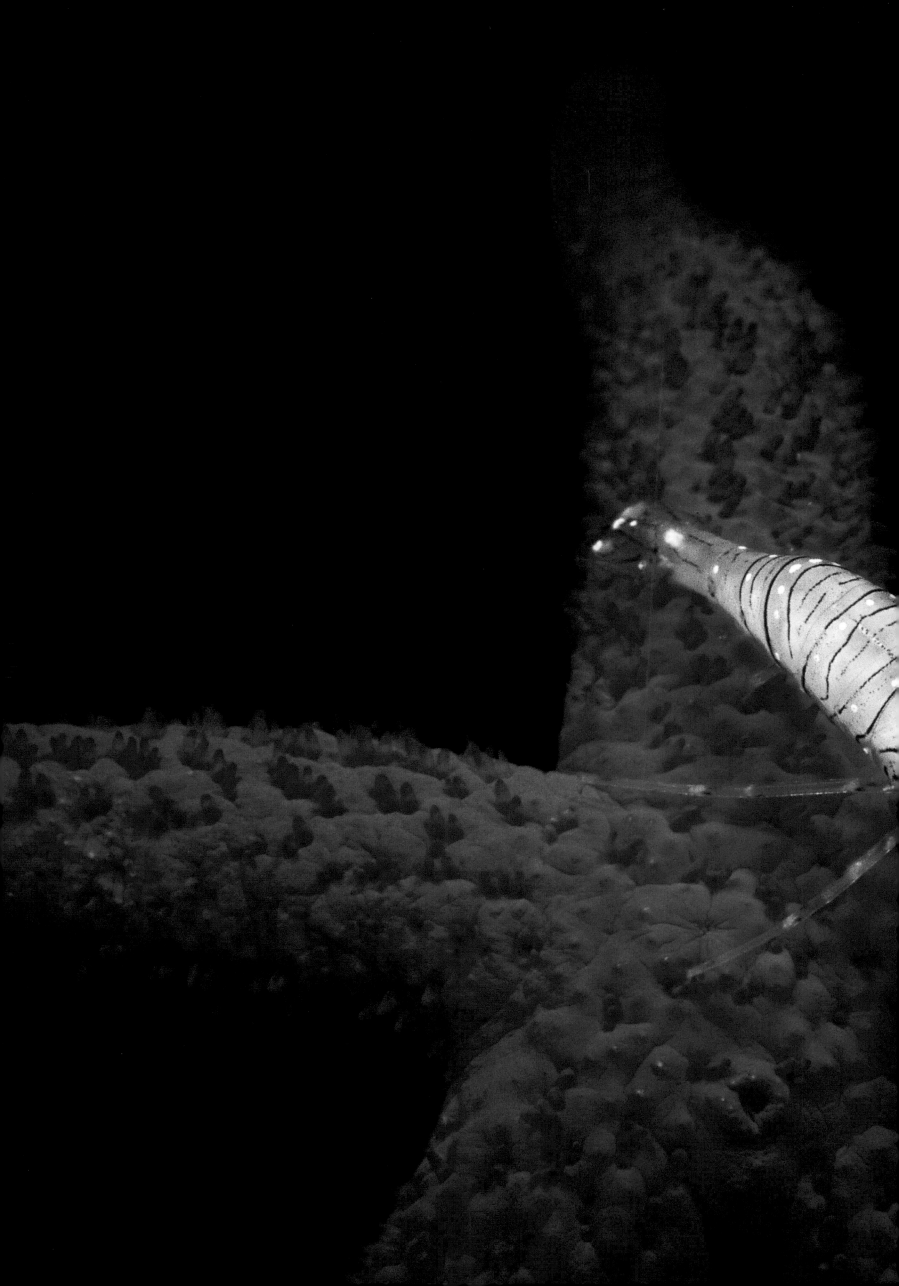

# CORSICA

The Mediterranean: posidonia,

*In the straits of Bonifacio, the Lavezzi Islands*

red coral and thick-lipped grouper:

*guard the secrets of Mare Nostrum,*

the sea of Homer's *Odyssey*.

*with mother-of-pearl and octopuses:*

*to the delight of the blue and white dolphin.*

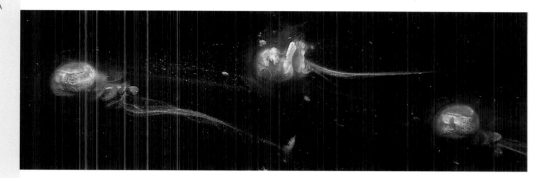

Corsica...the cliffs and inlet of Bonifacio... Is this the country of the Laestrygonians? Homer describes it thus in the Odyssey: 'On the seventh day, we reached the country of the Laestrygonians. We entered a port unknown to sailors. A steep double cliff rose all about us: ahead two long headlands faced us, constricting the entrance.'

A true to life scene... At Bonifacio Sophie wants to dive in the straits separating Corsica from Sardinia. She plans to reach the Lavezzi (Lavezzu) Islands, where there are reefs (Porraggia, Ratino, Piana, Sperduto...) flanking two of the largest — Cavallu and Lavezzu. On the Isle of Beauty (Kallyste, the 'Most Beautiful' of the ancient Greeks), she has already visited the seabed at Calvi, the Scandola marine reserve and the Bonifacio caves: at Sdragonato there is a bed of pink calcareous algae: the cave at Saint-Antoine, nicknamed Napoleon's Hat after its shape, bristles with stalactites. These caves give up their essential secrets, those known to the blue and white dolphin. There are some particular species that seem to have escaped from some other geological era and have an existence difficult for us to comprehend: this is exactly what Sophie is searching for!...This is the kingdom where the sea crow (*Sciaena umbra*) in its attractive dusky livery, the brown grouper (*Polyprion americanum*) and even the strange humantin shark (sea-pig) *Oxynotus centrina*) appear too recent. This place seems to have been handed down to sponges, anemones, worms, crustaceans, molluscs and echinoderms — in short, to the invertebrates of ages past. To the red coral, the famous *Coralium rubrum*, this precious cousin to the gorgonias (with eight-tentacled polyps), sought out since antiquity, loved too much, plundered too much, now rare, thus ever more precious.

## THE SECRETS OF THE *MARE NOSTRUM*

The Mediterranean still guards its mysteries well. The enigma of its population had not been solved until recently. Some 6 million years ago, the Straits of Gibraltar were closed. Cut off from the Atlantic and inadequately fed by its rivers, the basin evaporated. When emptied, nothing remained in the bottom of this 6,000 kilometre (3,700 mile) canyon except some strings and lakes of concentrated brine. A million years later tectonic forces re-opened the Straits of Gibraltar. The waters of the Atlantic rushed through the breech of the Pillars of Hercules. A prodigious waterfall, equivalent to 80 Amazon rivers, poured into the basin, taking a thousand years to fill it! The result? Three-quarters of the plants and animals of the Mediterranean come from the near Atlantic.

The Mediterranean... Sophie has dived and photographed it a little everywhere, the basin of her childhood from Cadaques in Spain, to the Marseilles Islands, to Corsica, the straits of Messina... Seeing the birds on the Bonifacio cliffs she dreams of

PRECEDING DOUBLE-PAGE SPREAD
The pink shrimp or prawn *(Palaemon = Leander serratus)* lies in ambush on an orange starfish *(Echinaster sepositus).*

Length 6 cm (2.4 in)
Depth: 25 m (80 ft)
Site: Lerin Islands

OPPOSITE PAGE, ABOVE
The phosphorescent pelagic mauve stinger jellyfish *(Pelagia noctiluca)* shines in the dark.

Length: 20 cm (8 in)
Depth: 25 m (80 ft)
Site: Cape Lardier

OPPOSITE PAGE, BELOW
The sea hare or aplysia *(Aplysia depilans)* is a large snail with an atrophied internal shell; here it is seen on a clump of forked algae.

Length: 20 cm (8 in)
Depth: 18 m (60 ft)
Site: Messina, Viola Coast

ABOVE
These strange, luminous chains are composed of tiny jellyfish colonies.

Length: 5 cm (2 in)
Depth: 5 m (16 ft)
Site: Lavezzi Islands

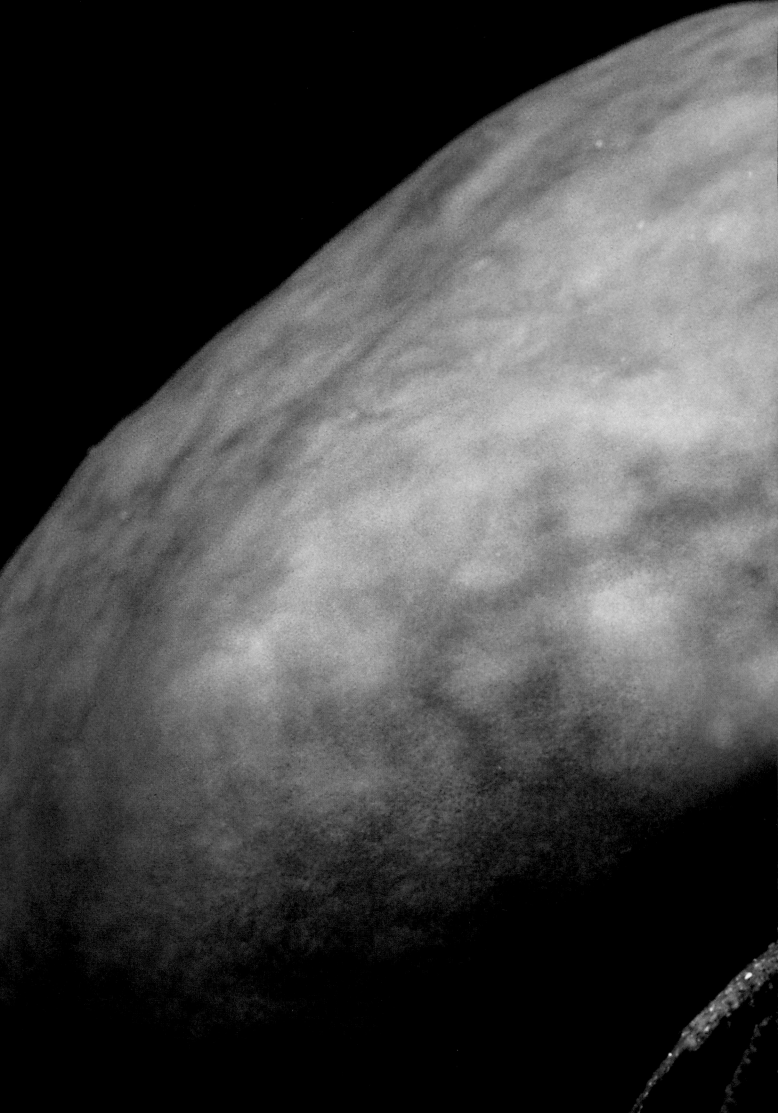

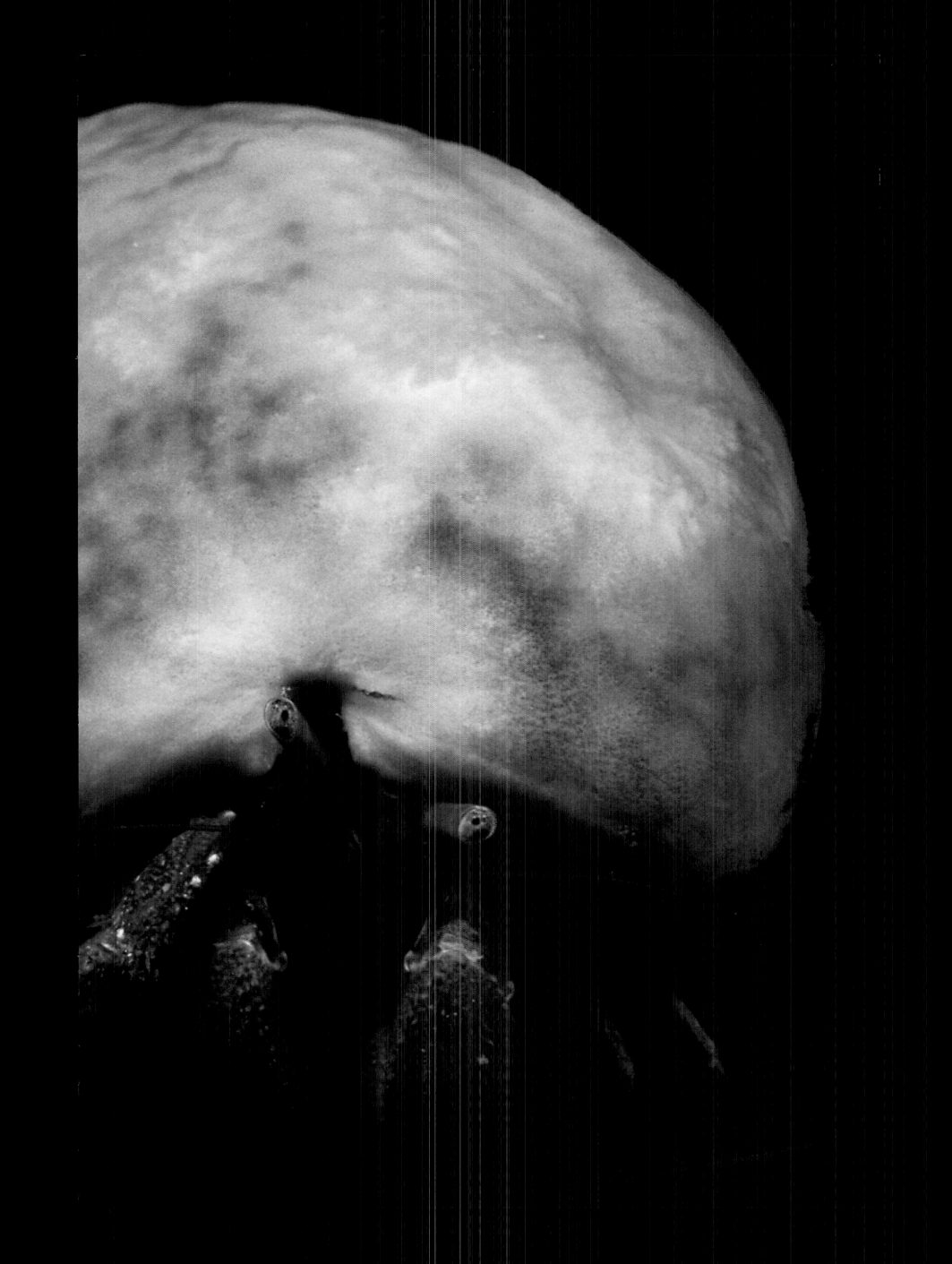

ABOVE
The Mediterranean
moray (Muraena helena)
appears menacing.
The species is
peaceful and
non-poisonous,
but, if it is aroused
and bites, can cause
a serious infection.

Length: 1 m (3.3 ft)
Depth: 40 m (130 ft)
Site: Scilla

the wind out at sea: here are the Caspian gull (yellow-beaked) and the rare (red-beaked) Audoin's gull, the black-headed gull, the peregrine falcon, Cory's shearwater, the petrel fulmar and the shag: not to mention the osprey diving to take a fish. Sophie drifts towards Lavezzu Island whose round, white rocks, or *cantoni*, harbour life's treasures above and below the surface. She contemplates the tragic reef on which the ship *La Semillante* was wrecked in 1855, since then called the 'Semillante Pyramid'. Should we not fear the risk that one stormy day one of the numerous tankers that ply this route may be wrecked on a headland and cause a disaster?

## POSEIDON'S GRASSES

It is time to experience the Big Blue. Sophie floats in the waves. She plays with some monk seals (*Monachus monachus*) — those elegant

Mediterranean pinnipeds, described by Aristotle in his *History of Animals*, that have been brought to the verge of extinction by centuries of human hatred. Now the species can count no more than three hundred and fifty individuals. The Lavezzu Islands were classified as a nature reserve in 1982. Some stupid people still poach dorado, mullet and grouper there, but the fish are returning.

The diver explores a field of posidonias (*Posidonia oceanica*), those marine plants dedicated to the god Poseidon. She threads her way through a jungle of clumps or 'mattes' of bottle-green ribbon-like leaves something like leeks, grazed by armies of violet sea urchins and salps, silver fish with longitudinal pale green stripes. She tries to come across the rare yellow flowering of this plant, which is not an alga but a higher plant — an angiosperm-bearing the brown ovoid fruits sometimes known as 'sea olives'.

From the surface down to a depth of 45 metres (150 ft), the posidonia beds form the basis of the ecosystems in the Mediterranean: they act as food-store, lung and nursery... Sophie picks out the algae between the 'mattes': the fan-shaped peacock's tail; the delicate mermaid's wineglass (*Acetabularia acetabulum*) shaped like a striped cap mushroom, 5 centimetres (2 inches) long but formed from a single cell; the cystoceira and forked dictyota; red algae tinged with pink, mauve or crimson... The Mediterranean clone with its yew-like leaves (*Caulerpa taxifolia*), imported from the tropics by humans (apparently via Monaco Aquarium), which is establishing itself in the Mediterranean at the expense of indigenous species, does not seem to have reached this far. A beautiful yet disquieting invader, in a green that is almost fluorescent... Sophie descends and is lost between the rocks. She recognises clathrinid sponges dressed alluringly in crumpled pale yellow chiffon. She glides over a crowd of bushy red and gold paramuricea gorgonia (*Paramuricea chamaleon*). She notes the pale, thin, pink sea fan (*Eunicella verrucosa*), sea-pens, sea-feathers, cerianths, anemones... She surprises Spallanzani's spiral tubeworm (*Spirograbnis spallanzanii*) spreading its plume of ochre and white tentacles. She finds a strange green spoonworm (*Bonellia uiridis*), whose female, equipped with a flexible 'horn' and a large rounded 'belly', is a metre (3.3 feet) long, while the male measures no more than a centimetre (0.4 inch). Here are molluscs: thorny oyster, sea snail and great scallop; the enormous fan mussel (*Pinna nobilis*), that reaches a height of 90 centimetres (36 inches), yawning in the current; the rare, supreme Mediterranean pear cowrie; murex; the triton (*Tritonium nodiferum*), a huge snail with a 40-centimetre (16-inch) shell.

ABOVE
The black scorpion fish *(Scorpaena porcus)* with its bristling, tattered skin can easily conceal itself among the rocks. The fins bear poisonous spines.

Length: 20 cm (8 in)
Depth: 45 m (150 ft)
Site: Calvi

The squid and the cuttlefish are there too, of course: these chameleons of the sea change their colours so quickly that they give the impression of setting fireworks off on their bodies. In a hole, Sophie discovers the gold, black-barred eye of an octopus (*Octopus vulgaris*). It is a female. She has laid hundreds of white eggs in strings, which she has stuck to the ceiling. She oxygenates and defends these promises of life without food or rest. Once they hatch she will die, exhausted.

## PLAYING BEST FRIENDS WITH FISH

Sophie scrapes the seabed. She photographs the delicate periclimenes shrimp, the velvet crab with its spiny shell, the crayfish, the box crab...

She finds the orange starfish and the ochre sand star. She observes a sea cucumber, like some obscene sausage, emitting a stream of sexual material in long white jets. She greets the gorgon's head brittle-star (gorgonocephalus) its twisting, divided arms forming a mane. She admires the transparency of the ascidians such as the crystal bell clavelinid.
On the gorgonias are nudibranchs, startling sea-slugs bristling with glinting electric fingers. The facelinas (*Facelina drummondi*) exhibit their sky blue appendages. The brown-spotted beige peltodoris sea-slug crawls by. Pink and mauve flabellinas (*Flabellina affinis*) execute a hermaphroditic ballet of love. The photographer sinks beside a wall. She plays friendly with the crowds of fishes. Male wrasse

(green, cuckoo and Mediterranean) make their nests on the sandy ledges. The royal rainbow wrasse (*Coris julis*) are dappled in orange and green. Each begins as a drab female being — its 'giofredi' stage; then it acquires male gender and glorious coloration: this is the 'julis' form. The chromis damsel fish shoot out a hail of violet-black. The damsels are orange-pink, the painted comber daubed in cabalistic signs... Here snakes a young conger, a moustachioed rockling, a spotted dogfish attaching its rectangular egg-case to a gorgonian. White sea bream, sand smelt, blotched picard, dorado, all on their way through life. An oreo dory with a big black spot on each flank. An almost circular sunfish (*Mola mola*) passes, moon-silver like the goddess Selene. Sophie is intoxicated with beauty. She rocks her head to imitate the Mediterranean moray (*Muraena helena*), its muzzle protruding from what might be the neck of a Roman amphora 20 centuries old. A cave... the diver creeps into the dark. The ceiling is carpeted with branches of red coral holding out their white polyps to catch plankton. A delta-shaped crevice is home to a dusky grouper (*Epinephelus guaza*). This is a big one, therefore a male! For the Mediterranean king grouper is a transsexual too. It starts its life as a female. At the age of about twelve, the masculine part of its genitals takes over. It ceases laying eggs and instead produces sperm or milt. This change depends on the needs of the species — the sex ratio of the group. When a reproductive male disappears, a female dedicates itself and matures its testicles to replace it. Lord Grouper can live, philosophically, for more than half a century. Sophie approaches. The fish opens its great mouth and 'bang!' an explosion... By suddenly opening its mouth the animal causes a cavitation effect and creates a shock wave. Dissuasion is guaranteed. The photographer removes her mask for a moment and grins like a child.

*The royal Mediterranean grouper starts life as a female and finishes it as a male...*

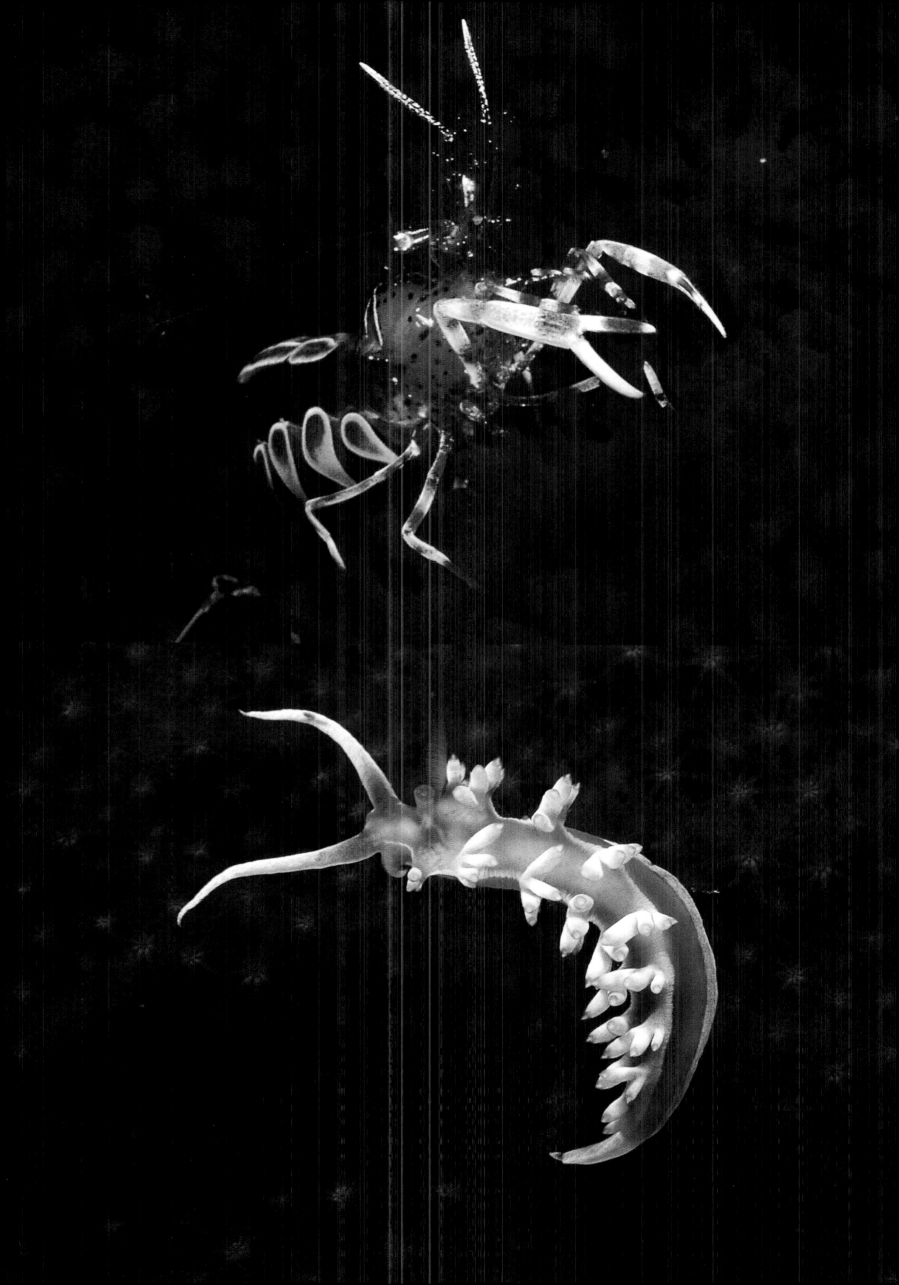

# THE SINAI

The spirit of the Eternal seems

*At Cape Ras Mohammed, the*

to breathe here in the Red Sea,

*magnificence shared by the desert*

between the Gulf of Suez

*and the sea creates a harmony*

and the Gulf of Aqaba

*between beings and beliefs.*

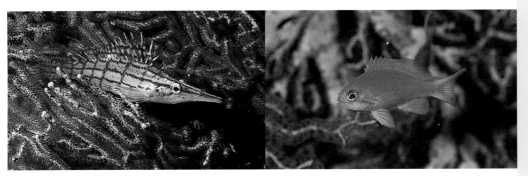

At dawn, the combined yellow and grey of the mineral fabric of the Sinai is suffused with a red and gold glow... At the end of the peninsula, planted in the blue of the Red Sea, there rises a headland called 'Ras Mohammed' — 'Cape Mohammed'. Moses and the Tables of the Law. Mohammed leaving for Medina at the start of the Hejira: it is difficult to avoid religious themes... This is the land of Islam, but one has the impression that the biblical chant of the Exodus still rises among the corals that pulse beneath the surface. But the Red Sea was more than just the sea that was opened up by Yahweh for the Jews fleeing from Egypt. That hardly serves as an adequate description of an expanse crossed by the Musulman to reach Mecca. Evolution chose this place to demonstrate the stupefying variety of the forms and colours of life. Even for those who have no beliefs — neither in the Bible nor the Koran — will pause to wonder. Between the Gulf of Suez to the west and the Gulf of Aqaba to the east, the coral colonies make up one of the most fabulous sights it is possible to find while diving. In these parts, the humble clown fish and the modest nudibranch are the angels in the paradise of Yahweh, God the Father, or Allah!

### AN IMPROBABLE UNIVERSE

Sophie contemplates this paradox of a landscape. On one hand, *terra firma*, the desert, a mineral empire: occasional acacias tortured into a bonsai existence, a stunted date palm, some spiny bushes in the less arid hollows; lizards, yellow scorpions and horned vipers. Some metres away, beneath the water, a madness of form and colour; the odds and ends of life; a paradise that is a riot of colours and geometries, of spirals and stars, lozenges, hexagons and spheres; cones, fringes and parabolas; flourishings and fluorescences, scenes painted by Hieronymus Bosch, Joachim Patenier or Goya.

Sophie readies herself. The sun burns her skin: there are no shadows on the shoals. The boat traces a slalom between the heads of coral. The Red Sea seems bluer than imaginable. Further out to sea teem the pink plankton shrimps that have given it its name. The coastal seabed is a palace in *A Thousand and One Nights*. A territory created by a magician... Here we can find fairy-tale beings with antennae (called 'crayfish'), with tentacles ('called octopus') with shells (murex, cowrie, triton, coneshell), with spiny carapaces (sea urchins and starfish), and with serpentine bodies (congers and morays).

Sophie immerses herself in the clear blue water. A shoal of fish welcomes her. The batfish with its flattened body; from the front it is a knife blade; in profile, a triangle with convex sides and rounded corners (non-Euclidian geometry). This is *Platax orbicularis*... Between two pillars of coral swim some pink apogon, green damsel fish, and moontail bullseyes (*Priacanthus hamrur*), some yellow-tailed red mullet, black with white striped soap fish (*Grammistes*

---

PRECEDING
DOUBLE-PAGE SPREAD
The transparent whip goby (*Bryaninops ampulus*) rests motionless on a scarlet gorgonia (*Juncella rubra*). Hard to find!

Length: 5 cm (2 in)
Depth: 35 m (115 ft)
Site: Ras um Sid

ABOVE, LEFT
The red 'wire netting' of this longnose hawk fish (*Oxycirrhites typus*) helps it to hide among the gorgonias.

Length: 10 cm (4 in)
Depth: 20 m (65 ft)
Site: Sharm el Sheikh

ABOVE, RIGHT
The little sea goldie (*Pseudanthias squamipinnis*) sinks into its gorgonia retreat.

Length: 12 cm (5 in)
Depth: 30 m (100 ft)
Site: Ras Mohammed

OPPOSITE PAGE
This colony of encrusting sea squirt (*Eusynstyela latericius*) belongs in a hallucination. Together, its members form a kind of 'space-saving' mosaic on the substrate.

Length: 0.5 cm (0.2 in)
Depth: 30 m (100 ft)
Site: Seven Brothers

*sexlineatus*), red soldier fish, and the thicklip (*Plectorhyncus gaterinus*) with blue lines and violet dots.

On the reef drop-off, the magic turns to ardour. It becomes an urgency, a necessity. As if, in this environment, nothing normal could ever be suitable. As if reality itself were unreal... Sophie glides over acropore, montipore, mushroom and dendrophyll corals. She rounds a scrub of pink or mauve gorgonias and alcyonarians where pass dragon fish, three-spotted damsel fish, violet and white peacock grouper (*Cephalopholis argus*) and the marvellous crimson diamond grouper (*Cephalopholis miniata*) dappled with blue eyespots.

She remembers... when she was small, she knew Commander Cousteau. She was entranced by the first underwater scenes she saw, in his film *The Silent*

opposites: Beauty and Life!' Beauty and Life, the results of violence and death... Here, everything confirms this paradoxical relation.

Sophie lets herself drift into the transparent. 20 metres, 30 metres... (65 feet, 100 feet...). Suddenly the Ras Mohammed wreck rises up before her eyes. A colossal metallic structure. A blue-black monster against the royal blue water... The hull is starred with ten thousand invertebrates: tube sponges in casks or in crusts; orange and gold hydras and zoanthers (flower-animals), delicate sea-pens, gorgonias and alcyonarians; thorny and zigzag oysters, feather stars in black and silver-barbed corollas; precious necklaces of jade-tinted amethyst or topaz ascidians...

The beauty of the detail enhances that of the whole. The Ras Mohammed

*At the bottom of the sea, horror and death are transformed into their opposites: beauty and life!*

*World*... Some of the film was shot in the Red Sea, in fact several sequences were taken at Ras Mohammed. The crew of the *Calypso* had found a wreck, identified as the British cargo ship SS *Thistlegorm*, 10,000 tonnes, that had gone down during the Second World War, after a German air attack.

### THE COUSTEAU SOUVENIR

Cousteau's divers had explored the ship. A perfect wreck sitting upright on its keel in exceptionally clear water. Sophie had liked Cousteau's comment on this sunken ship: 'This is how the passions of war finish: in the magnificence of corals and gorgonias, the indifference of the grouper, the angel fish and soldier fish. At the bottom of the sea, horror and death are transformed into their

wreck fits in with the picture one has of a submerged ship when dreaming, at the cinema, or on a tape... It is not a shapeless mass, a pile of members and superstructure ruined by time, bacteria and boring animals; rather it is an intact glory; a legendary monument. This ship seems to dream of rising to the surface once more, returning to its old self, and sailing off into the infinite unknown.

A hump-head wrasse (*Cheilinus undulatus*), one of the two species of Napoleon fish of the southern seas (the other is a parrot fish) rises up behind a davit; it is hatched and lined in blue and green with, on its forehead, the hump that earned it nicknames, and the name 'lorry fish', donated by Cousteau's friends. 'Lorry' has not been usurped. This fish is more than 2 metres (7 feet)

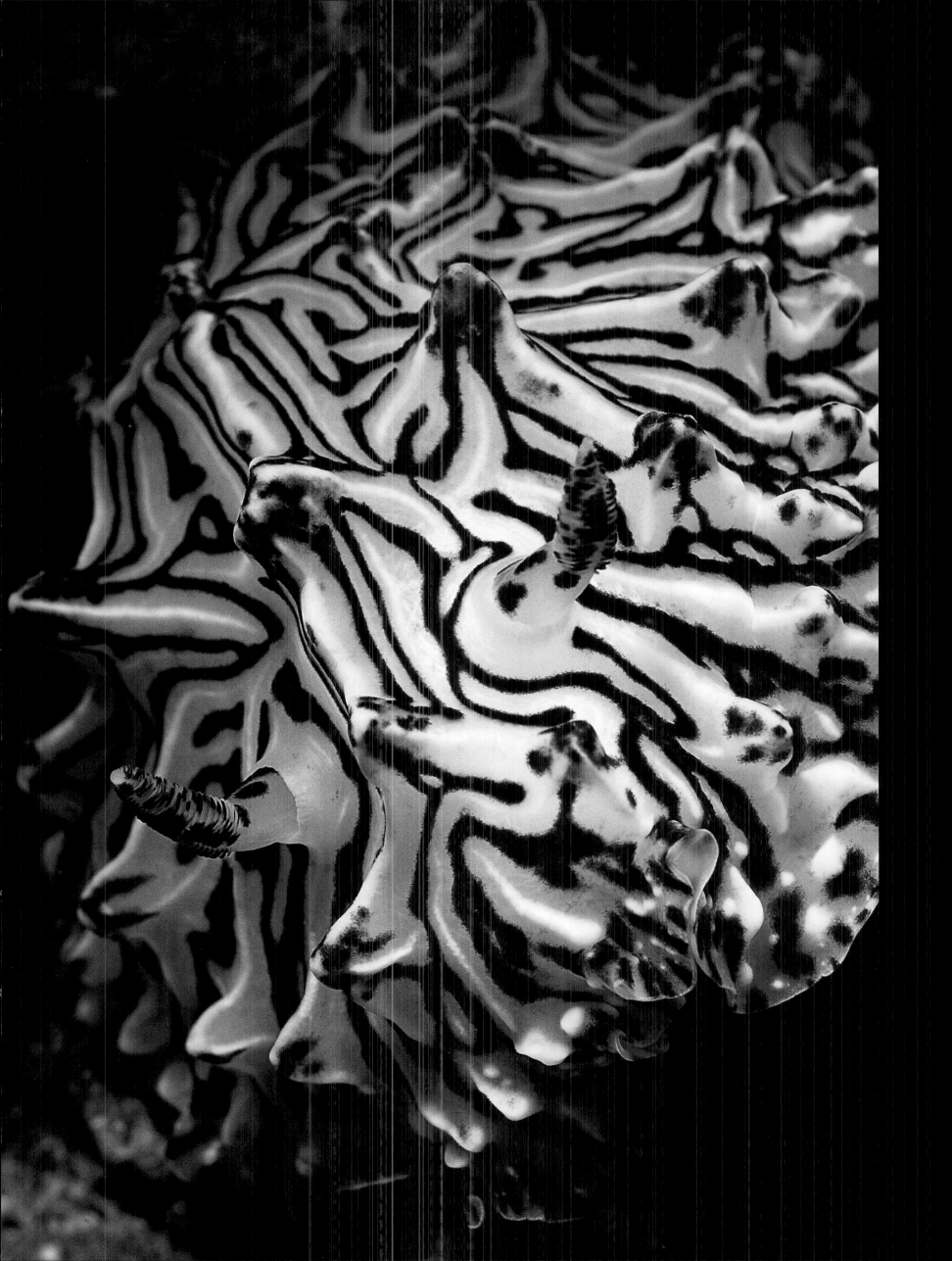

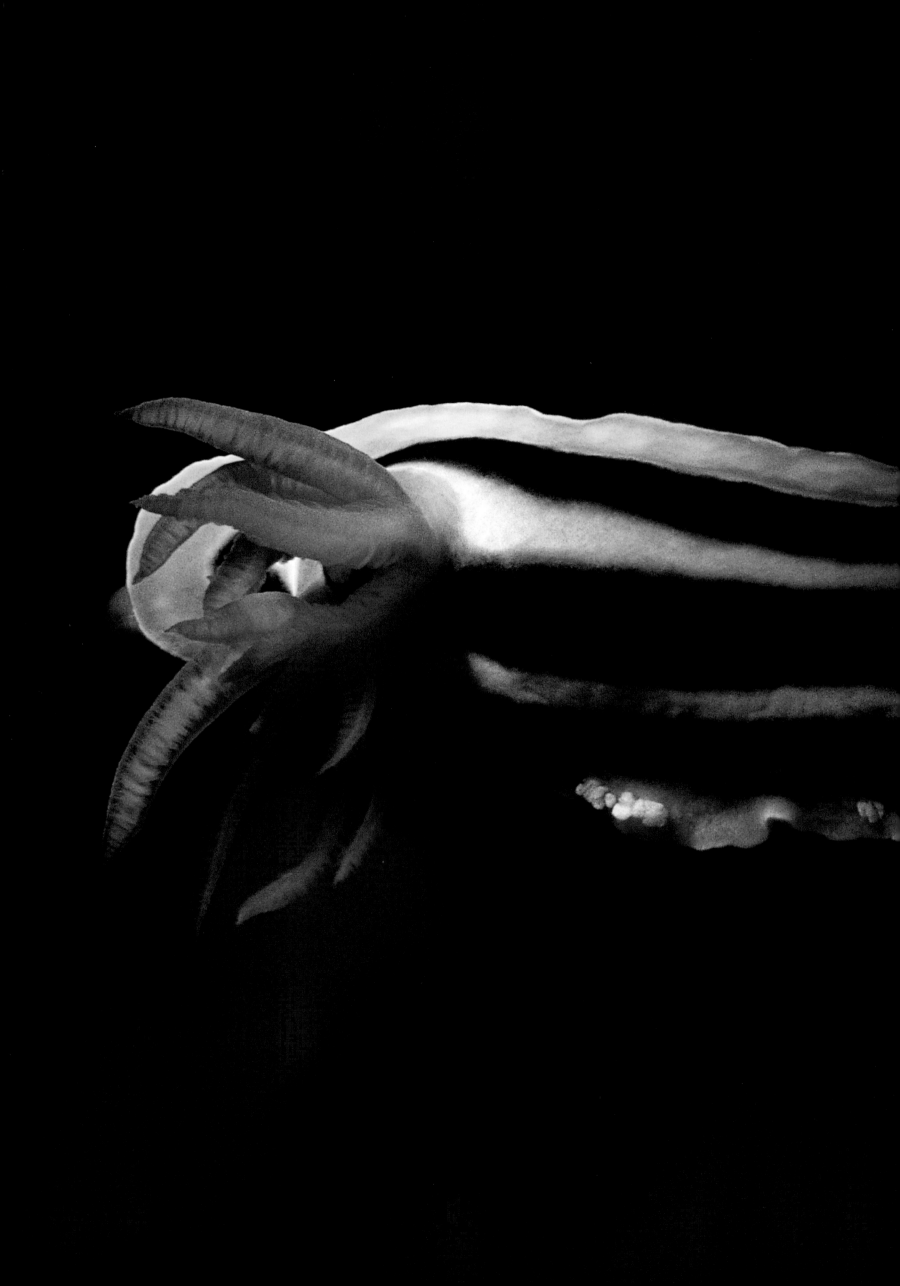

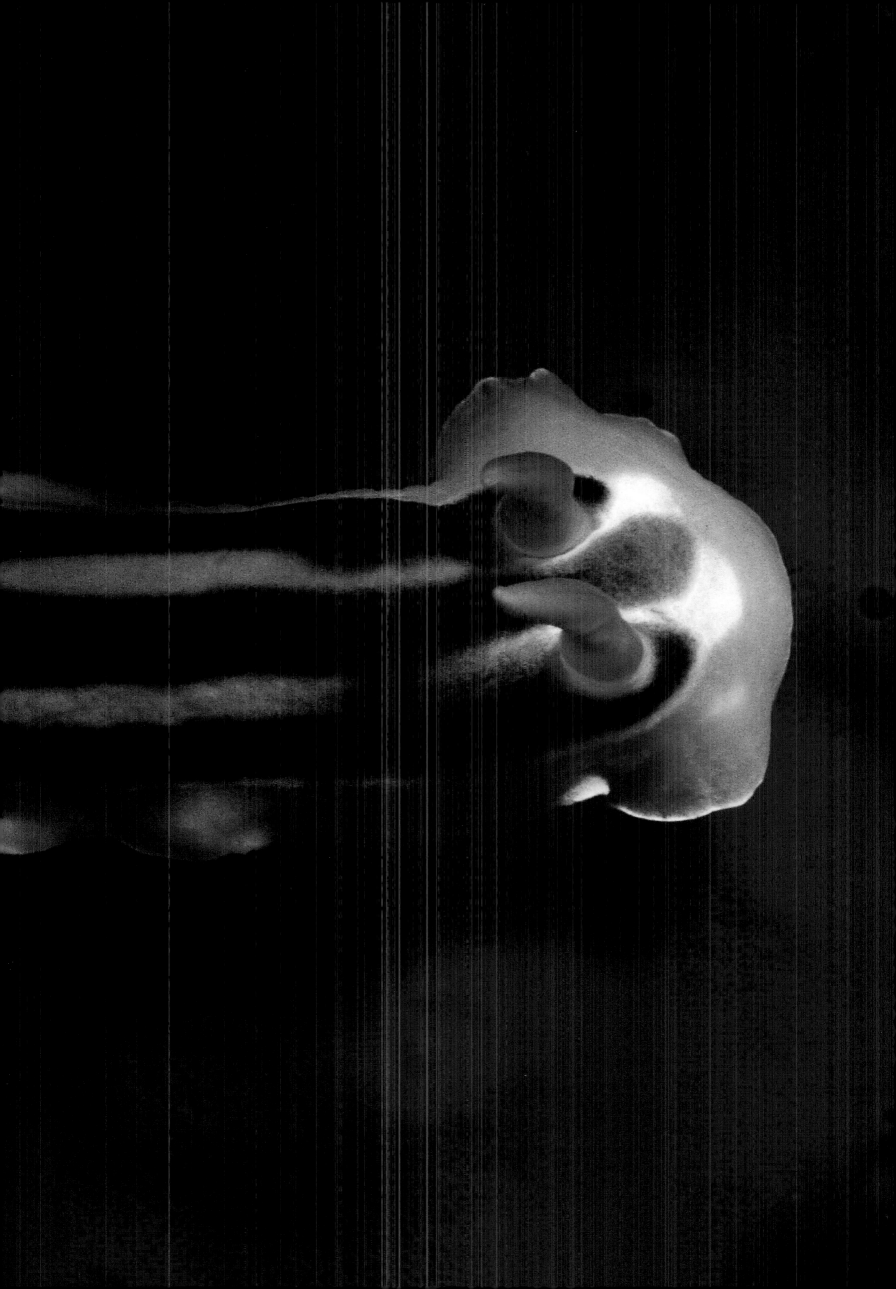

*All is splendour, pretence and oddity in this landscape rendered theatrical by luminous rays from the surface.*

long and must weigh 180 kilos (400 lbs). It approaches Sophie. It wants to know. It is curious. After all, it is at home and fears nobody.

The Napoleon fish comes to sniff at the stranger then, without excessive ceremony, goes about its business. Around about, amberjack embody predatorial efficiency and the height of the art of swimming. A white-tipped reef shark materialises in the fluid, followed by a couple of grey reef sharks; then three blacktip sharks. Sophie slips into a gangway bristling with pale wire gorgonias. She runs into a moray at the entrance to a hatch, its mouth gaping, its body hidden by the hull. A giant: *Gymnothorax javanicus*, brown-red with white spots, it exceeds 2 metres (7 feet). Its head is bigger than hers.

## SHA'AB RUMI AND THE SEVEN BROTHERS

There is no sign of an aggressive nature in this creature. The 'spitefulness' of morays is the invention of the spiteful diver who believes in his own fantasy. Sophie greets it eye to eye, an unusual form of the 'Sartres stare'. However, it is best to avoid provoking this creature and getting bitten; some morays have poisonous saliva; all have 'dirty' teeth, teeming with bacteria that can cause septicaemia or gangrene.

Sophie takes a turn around the Ras Mohammed wreck. A dozen spotted eagle rays (*Aetobatus narinari*) pass by, their tails like swords, beating their wings like pterodactyls. On a sandy expanse the diver finds a Moses sole (*Pardachirus marmoratus*), spotted in grey and brown, whose skins secretes the most effective shark-repellent substance known. All is splendid and unusual in this landscape rendered theatrical by luminous rays striking down from the surface. In the folds of a leaf coral, a tiny King Solomon fish or orchid dottyback (*Pseudochromis fridmani*) crackles with violet-pink. Amid the long spines of a diadem sea urchin lodge dozens of striped apogons.

Enthroned on a rock, a reef stone-fish (*Synanceia verrucosa*) is so perfectly camouflaged as to be almost invisible. Its dorsal spines can inject those foolish enough to touch them with one of the most virulent poisons in the animal kingdom.

The Ras Mohammed adventure is just the first of a series in the Red Sea. The diver is to visit the reefs of Sharm el Sheikh and Ras Nasrani (Egypt), of Eilat (Israel) and Aqaba (Jordan) at the end of the gulf of the same name. Further south, in the Sudan near Port Sudan, she will cover the seabeds of Ras um Sid, Sha'ab Suedi and Sanganeb, then those of Sha'ab Rumi where, in 1963, Commander Cousteau ran the submarine house *Precontinent II*. She will thread her way toward the Farasan Islands close to Saudi Arabia. She will explore the reefs of the Seven Brothers (or Seven Apostles) near Djibouti, and those of the strait of Bab el Mandeb, which separates the Red Sea from the Gulf of Aden and the Indian Ocean. She will follow in the tracks of Henry de Monfried, but under the water, in the country of sea-lilies and gorgonias. In the intimacy shared by sea hares and comb jellies as white as angels, in the paradise of all religions… The young woman smiles at the sight of a bottlenose dolphin surfacing from the liquid mass. The cetacean leaves a necklace of bubbles, clicking a message in the language of the ocean. What human can understand?

**PRECEDING DOUBLE-PAGE SPREAD**
A striped chromodoris sea-slug (*Chromodoris elizabethina*) is among the most striking of nudibranchs. It declares 'I am poisonous!' in glorious colour.

Length: 4 cm (1.6 in)
Depth: 15 m (50 ft)
Site: Ras Nasrani

**ABOVE**
'Who am I?' Probably a dwarf goby (*Bryaninops* sp.) a discreet, though essential tenant of the reef.

Length: 2 cm (0.8 in)
Depth: 20 m (65 ft)
Site: Seven Brothers

**OPPOSITE PAGE**
The coral oyster, a pectinid (*Pedum spondylium*) just manages to open in a mass of the boring sponge (*Clionia vastifica*).

Length: 5 cm (2 in)
Depth: 15 m (50 ft)
Site: Sanganeb

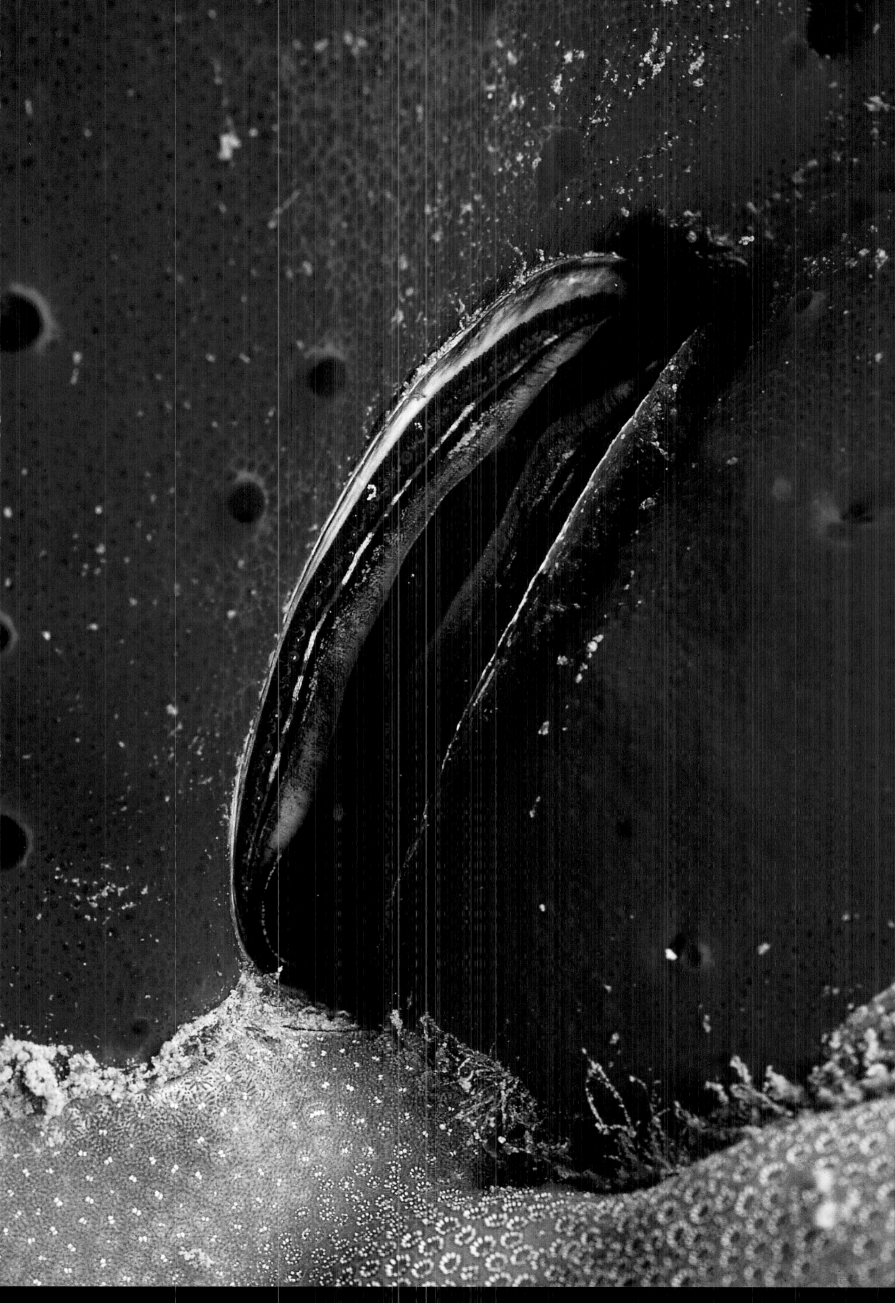

# THE SEYCHELLES

## Aldabra, the most

*The most original and the best preserved.*

## secret island in

*Where the slow pace of giant turtles*

## the archipelago.

*echoes in the ocean depths.*

# THE SEYCHELLES

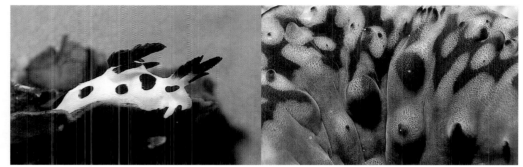

Aldabra… Sophie sits close to a giant turtle. The reptile lugs its years through the bushes. An antediluvian head, wrinkled neck and rounded shell: one would say that time had stopped for it, as it did in the Galapagos, far from here. Giant turtles have been crawling towards their destiny in these archipelagos since the age of the dinosaurs. The patience of evolution!

Magnificent frigate birds streak through the blue and, in full flight, snatch the fishy prize caught by their fellow aerial spirits — seagulls, terns, herring gulls, and the crimson finch (the 'tropical bird' beloved of poets). The white terns are nesting. But, for these swallows of the sea, nesting does not mean 'building a home'. The female lays an egg on a branch. The prospects of the egg and chick depend on a precarious balance through wind and storm. In this we see a symbol of life on earth: a frail, precious embryo, a fragile shell… Sophie makes herself ready on the shore, where tortoises and their marine cousins (the green, scaled, loggerhead turtles) cross paths, though they ignore each other: the females of these ocean nomads come to lay their eggs in the sand, leaving tank tracks on the beach. The young woman leaves the sand and enters the warm water, crossing a pale blue lagoon and some gently sloping shoals. This archipelago is a sort of absolute. A distance among distances, where the turtles' message is that the world is slow, but goes quickly.

Between fish, birds and man, fathom it out he who may!

The Seychelles throw their marvellous confetti over the Indian Ocean, off the coast of Africa, from the equator to near Madagascar and the Comoros. The northern group, including the Bird Islands, Denis, Praslin, La Digue, Felicity, Silhouette, the Frigate Islands, etc., and Mahé (with Victoria, its capital), hosts most of the human population. This pattern of islands stretches south-west through the islands of Plate, Amirantes, Alphonse, Providence and Farquhar as far as the sentinel Cosmoledo, Assumption and Aldabra. Geology can produce surprises. In contrast with most coral archipelagos, there are no ancient volcanoes here, only granite peaks. The Seychelles are small and scattered. Their sand and coconut palm beaches remind one of Robinson Crusoe. (There is the ordinary coconut, and there is the coco-de-mer or 'backside' coconut, native to Praslin.) Together, the Seychelles amount to a continent. They are born with the slow movement of the earth's crust, the offspring of tectonic plates; of the great drift that separated Madagascar from the African continent.

## THE INDIFFERENCE OF THE MANTA RAY

Every underwater foray seems like a journey in time and space. Feelings experienced in the liquid universe are different from those in the free air.

PRECEDING
DOUBLE-PAGE SPREAD
The flamenco red of the nudibranch *Hexabranchus sanguineus*: the little Spanish dancer performs her dance of the flying veils over the reef.

Length: 30 cm (12 in)
Depth: 35 m (115 ft)
Site: Aldabra

OPPOSITE PAGE
The blue and yellow nudibranch *Phyllidia varicosa* ( = *P. arabica*) secretes such powerful toxins that it has no predators!

Length: 7 cm (2.8 in)
Depth: 25 m (80 ft)
Site: Mahé

ABOVE, LEFT
This polycera nudibranch (*Polycera sp.*) has never been described: an unlikely and memorable encounter.

Length: 2.5 cm (1 in)
Depth: 20 m (65 ft)
Site: Aldabra

ABOVE, RIGHT
A detail of the lines, excrescences and oddities of the skin of the nudibranch *Halgerda willeyi*.

Length: 1 cm (0.4 in)
Depth: 50 m (165 ft)
Site: Aldabra

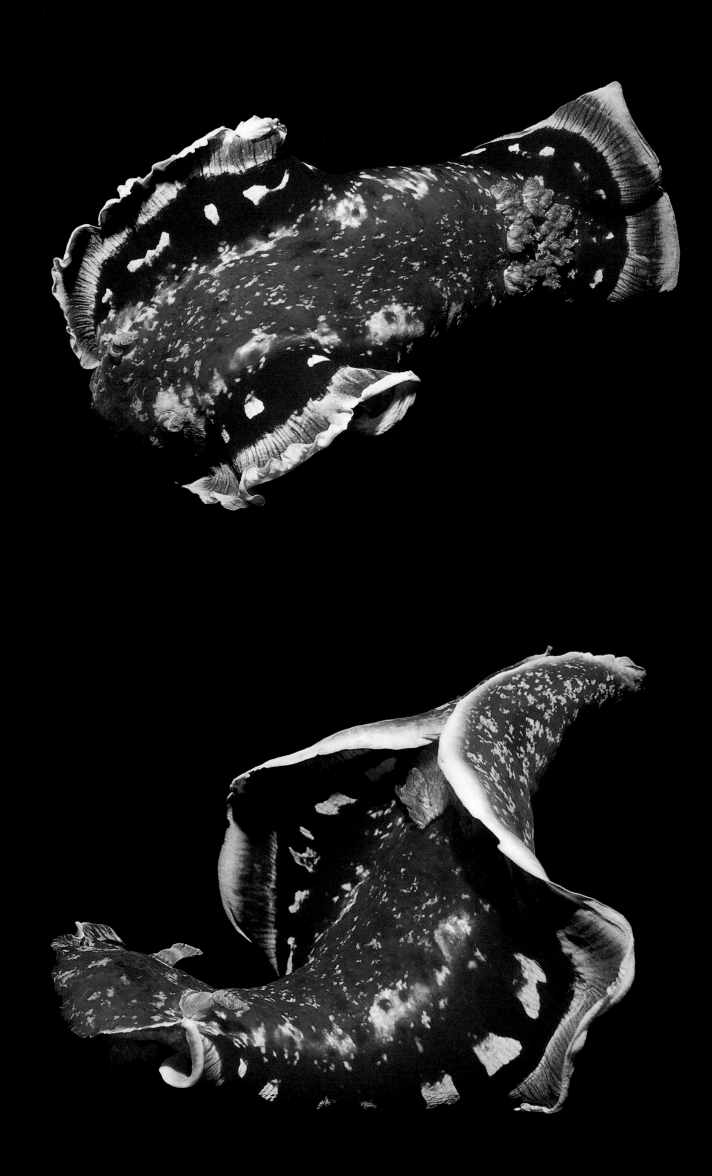

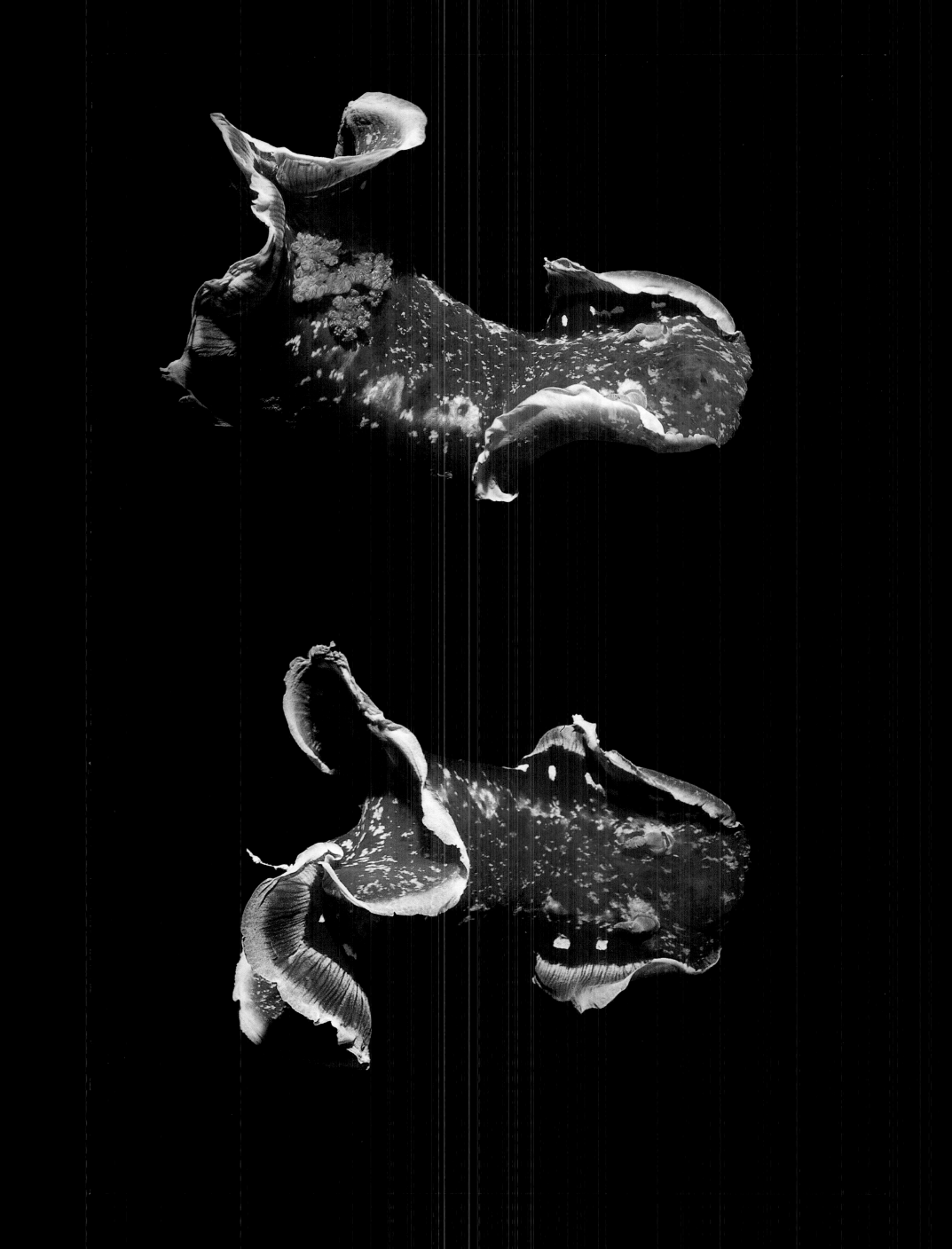

One not only becomes weightless, like an angel (or an angel fish…), but visual tonality changes (everything becomes more blue, then darker), sounds become distorted, and smells disappear. The brain adjusts its references. A kind of laziness sets in; harmless, unlike the depth narcosis due to the effect of excessive nitrogen on the nerves. You are overcome by a feeling

Princes among the 'small' animals are the nudibranchs, the sea-slugs. It is these that Sophie seeks. She swims over giant ochre *Subergorgia mollis*, then cream whip gorgonias and brick-red brush gorgonias. In a sandy bed she recognises sea-feathers and comb-shaped sea-pens stuck in the substrate by their handles. She examines the magnificent sea anemone (*Heteractis magnifica = Randianthus ritteri*), then its

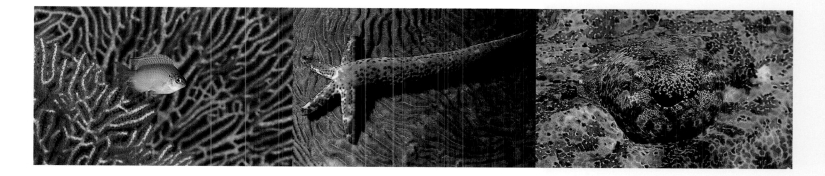

of well-being. The ocean is as soft as the amniotic fluid in which we bathe before birth. You become a foetus again, aware of the reassuring beat of the great mother's heart.

## SEA-FEATHERS AND GORGONIAS

Sophie floats along a coral wall. Green coins of halimed coralline algae and silvery peacock's tail fans colonise the sandy ledges while red lithothamnes algae spread their pink blades along the crest of the reef. Fields of ball, tube, encrusting and barrel sponges take on the most surprising hues: violet-green, brown-mauve, blue-orange. Hydras, tubipores and great pale *Sarcophyton* alcyonarians alternate with magic clumps of the red-pink or mauve *Dendronephthya*. In the open water swims a manta ray. One might dream up such a fish, with its arrow of a tail and its enormous triangular wings (white below, blue-black above). The two cephalic 'horns' that guide the plankton into its mouth have earned it the name 'sea devil'. To Sophie, the manta ray embodies an angel; one of the most beautiful 'large' animals in the ocean.

pearly cousin (*Heteractis aurora*), frequented by the white-striped orange clown fish (*Amphiprion clarkii*). A giant grey-green *Sticnodactyla* hosts a pretty periclimenes shrimp and some other clown fish. The complicated architecture of the true corals asserts itself: cushions of needle coral and pink-fingered stylophora; leafy bottle-green montipora and acropora (*Acropora humilis*); poritidae and goniopora, mushroom corals and lettuce coral, bubble coral, scroll coral… *Pseudoceros* flatworms show off their gay, varied colouring: these scraps of ribbon — striped, banded, sinuous, punctured, inlaid — undulate across the seabed. The annelid worms are also represented here: by giant spirobranchs, their double plumes of tentacles showing like white, blue, red or yellow pompoms, ready to withdraw into their tubes at the slightest alarm. Of course there are the molluscs too. Sophie notices bivalves such as the hinged thorny oyster, a row of blue eyes at the edge of its mantle; and here a painted pearl oyster (*Pinctada margaritifera*): it is impossible to imagine it might conceal a treasure. Yawning, this huge fan mussel reveals

PRECEDING DOUBLE-PAGE SPREAD
Four poses by the little Spanish dancer *(Hexabranchus sanguineus)*: this large sea-slug swims by flexing its flat, fringed body.

Length: 30 cm (12 in)
Depth: 20 m (65 ft)
Site: Aldabra

OPPOSITE PAGE
The spindle-shaped egg cowrie *(Hiata depressa)* merges with the branches of the orange gorgonian on which it lives, and whose polyps it readily eats.

Length: 1.5 cm (0.6 in)
Depth: 35 m (115 ft)
Site: Aldabra

ABOVE, LEFT
A delicate green damsel fish *(Chromis viridis)* searches the gorgonia for scraps to nibble.

Length: 8 cm (3 in)
Depth: 10 m (33 ft)
Site: Aldabra

ABOVE, CENTRE
The comet starfish *(Linckia multiflora)* can regenerate its entire being from a single arm.

Length: 8 cm (3 in)
Depth: 30 m (100 ft)
Site: Aldabra

ABOVE, RIGHT
The camouflaged eye of a crocodile fish *(Platycephalus longiceps)*… The branched cirri cover half the cornea, hiding the organ from the enemy's sight.

Length of the eye: 2 cm (0.8 in)
Depth: 35 m (115 ft)
Site Aldabra

the blue and green ripples of a princely mantle within. Farther on, in a z-shaped cleft, cuttlefish and squid vibrate their fins and display fireworks on their skins. Doubtless it is a language, but a language difficult for mere humans to understand... Snails abound: troches in pointed quack magician's hats; ovules, cowries and sea olives in various shades, patterns of colour and precious mottling; coneshells that can spit tiny venomous projectiles at the intruder; murex and spider conch, bristling with spikes... But the nudibranchs are the finest of all. The sea-slugs, long, flat molluscs that can weigh anything from 10 grams up to a kilogram (0.35 oz to 2.2 lb) and have no shell, though some bear calcareous needles in their skin. Having no true gills, they breathe by means of tufted

trick: they swallow tentacles from cnidarian coelenterates (corals and anemones); the poisonous cells from these then migrate, still alive, into the creatures' dorsal projections: from then on they are protected. There are two dozen families of these psychedelic creatures (notodores, polyceres, hexabranchs, chromodoris, facelina, etc.) that largely inhabit reefs in tropical seas. In Aldabra's waters there pass armadas of damsel fish and butterfly fish, surgeon fish, and trumpet fish with noses worthy of Cyrano; along with them blue parrot fish and armour-plated coffin fish, sergeant majors and unicorn fish, cow fish and sheep fish — that know no other pasture than algae and marine flora... A royal angel fish and an imperial angel fish — courtly cousins — rejoice in the splendour

*In Aldabra's waters pass armadas of damsel and butterfly fish.*

organs fore and aft on their bodies. These are called 'neoforms', perhaps because evolution only invented them, say, recently. A sluggish folly: the nudibranchs sport two pairs of tentacles on their heads, as well as all kinds of projections: knobs, points, swelling and dorsal expansions known as 'cerrata'. These are penetrated from within by hepatic diverticula, protrusions of the animal's liver. This anatomical peculiarity has long intrigued biologists, who have at least partly solved the mystery. These are chemical weapons of war. Sea-slugs are superbly painted because they are poisonous; the poison lies in their cerrata. By displaying their glorious colours, they are warning predators: 'Don't touch me — I'm inedible!' A number of species play another

of their yellow and blue stripes. They are determined to keep their star status among the corals. Sophie trains her camera lens on a warty phyllidian nudibranch *Phyllidia arabica*. It is about 8 centimetres (3 inches) long, with projections on the body, and speckled with gold on a background of grey-blue velvet set off by black lines... The nudibranch pulses gently, a mystery atop a sponge, crossing the path of a close relation, a four-coloured chromodoris *Chromodoris quadricolor*: 4 centimetres (1.6 inches) of sheer beauty, striped lengthwise with blue, white, black and yellow-orange. Like a re-invented flag of the Seychelles, waving over two shells like double coconuts and a grey-black sea cucumber, crawling on the seabed with all the gravity of a giant tortoise.

PRECEDING DOUBLE-PAGE SPREAD, LEFT
The tambja sea-slug (*Tambja luteolineata*) on its alga, resplendent in black and yellow lines picked out with jade.

Length: 3 cm (1.2 in)
Depth: 15 m (50 ft)
Site: Aldabra

PRECEDING DOUBLE-PAGE SPREAD, RIGHT
Bennett's sea-lily (*Oxycomanthus bennetti*), a crinoid, grasps an orange knotted gorgon (*Melithaea* sp.) with its numerous arms.

Length: 20 cm (8 in)
Depth: 40 m (130 ft)
Site: Mahé

THIS PAGE, LEFT
An as yet unidentified sea-slug, perhaps in the genus *Aegires*.

Length: 1 cm (0.4 in)
Depth: 25 m (80 ft)
Site: Aldabra

THIS PAGE, RIGHT
This helix is an antipatharian (*Cirripathes spiralis*), a black coral also known as 'spring' or 'wire' coral.

Length: 50 cm (20 in)
Depth: 50 m (165 ft)
Site: Aldabra

OPPOSITE PAGE
The majid sea-spider (*Picrocerus* sp.) camouflages itself by covering its body in debris.

Length: 4 cm (1.6 in)
Depth: 35 m (115 ft)
Site: Assumption Island

# THE MALDIVES

An archipelago that only

*Its coral islands lie so low that*

just exists... well away from

*in the twenty-first century they could*

everything — except splendour.

*disappear due to global warming.*

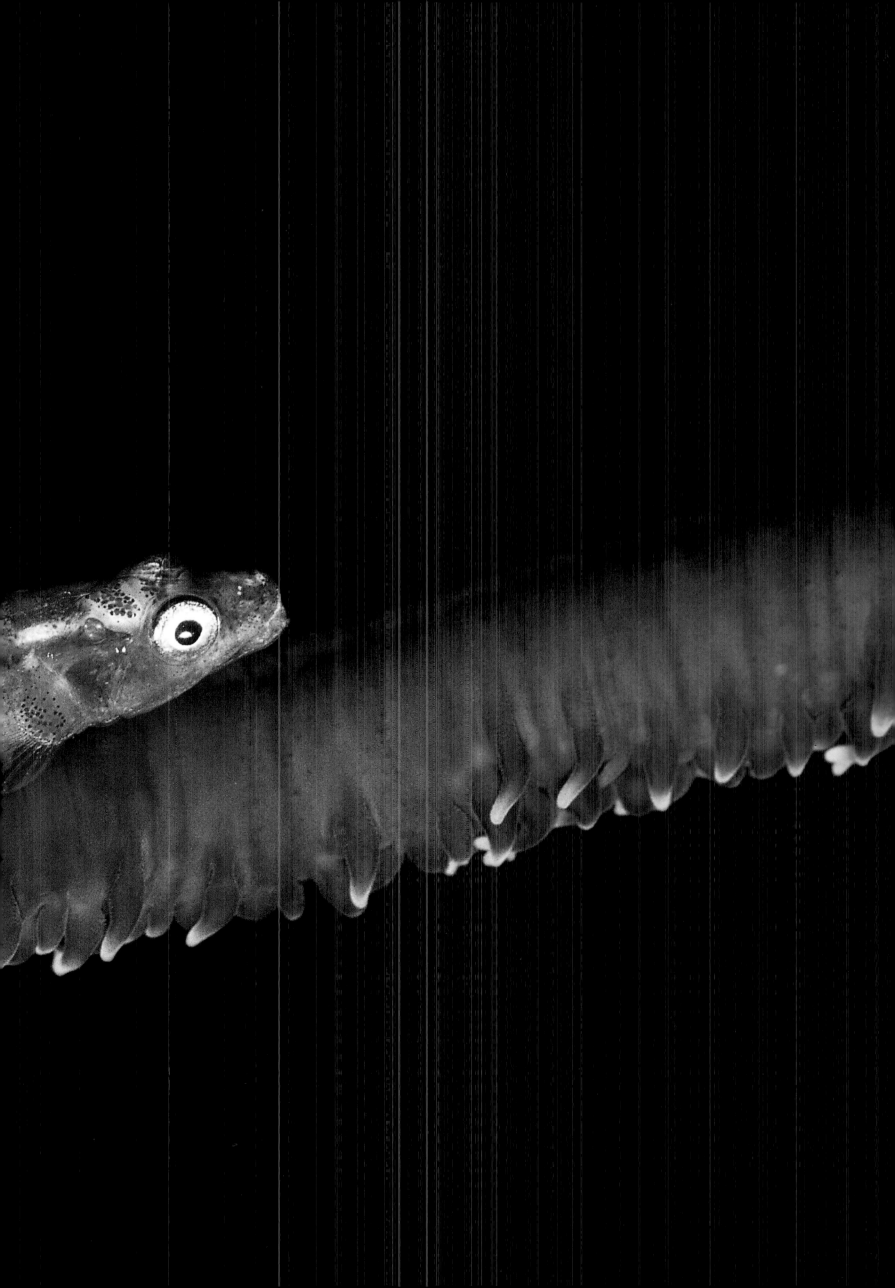

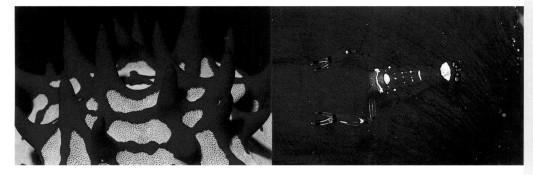

You can hardly make out the specks on a world map. However, this sprinkling of nothings on the intense blue of the Indian Ocean has a name: 'the Maldives'. A treasure under a suspended sentence... here is all that remains of a chain of submarine volcanoes born to a fissure in the earth's crust (a part of the mid-ocean ridge), some of which have risen from the sea, then submerged again, encrusted with a fringe of reefs and crowned by an atoll. The coconut palms sway... an illusion of perpetual peace... but will the Maldives exist at the end of the twenty-first century? Sophie is passionate about this archipelago. It was here, in 1984, after her father's death, that she first dived with a camera. It was in these waters that she created the best things in her life from a sadness mingled with joy; where she linked her pleasure at being in the world with the will to defend nature and beauty.

## A FORETASTE OF THE FLOOD

The Maldives make up a tiny state of nineteen main atolls and more than a thousand islands. Scarcely two hundred of these are inhabited. Total area: 298 square kilometres or 115 square miles; 170,000 inhabitants. Capital: Malé. In Maldivian, the local language, this confetti to the south-west of Sri Lanka is called *Maldvipa* (the Mal Islands). Separated from Laquedives by the Eighth Degree Channel, they begin at the Kelai atoll in the north and end beyond the equatorial channel (where so many merchant vessels from the Indies foundered), at the huge Addu atoll. Our photographer has already dived in the limpid waters of a few of these atolls. Besides Malé (and its companion Wadoo) she has visited Rangali, Rihiveli, Baros, Felidu... She has sailed for Ari atoll. The water is ultramarine, the sky azure with little puffs of cotton-wool clouds; the peace of the tropics pervades. At least for the moment, for the reefs are battered by monsoon storms, and sometimes by typhoons of alarming power... The reason why the islands risk submersion is that, because of the greenhouse effect, global warming is raising the level of the sea. These coral outcrops make up the first country in the twenty-first century that could meet the fate, which, according to myth, overtook Atlantis and the city of Ys. An inkling, a foretaste of the Flood...

No doubt Sophie has all of this in mind as she prepares her mask and camera, and submerges. Once more she remembers her first assignment in the Maldives; her rapture, which, for want of a better phrase, she described as her 'state of grace'. She lets herself drift in the friendly water. A large aurelia jellyfish armed with a fringe of short tentacles and with a four-leaf clover (its reproductive organs) at its centre, pulses its way between the pillars of coral that the ships of the East India company would have hated so. Here are branched acropora and leafy

---

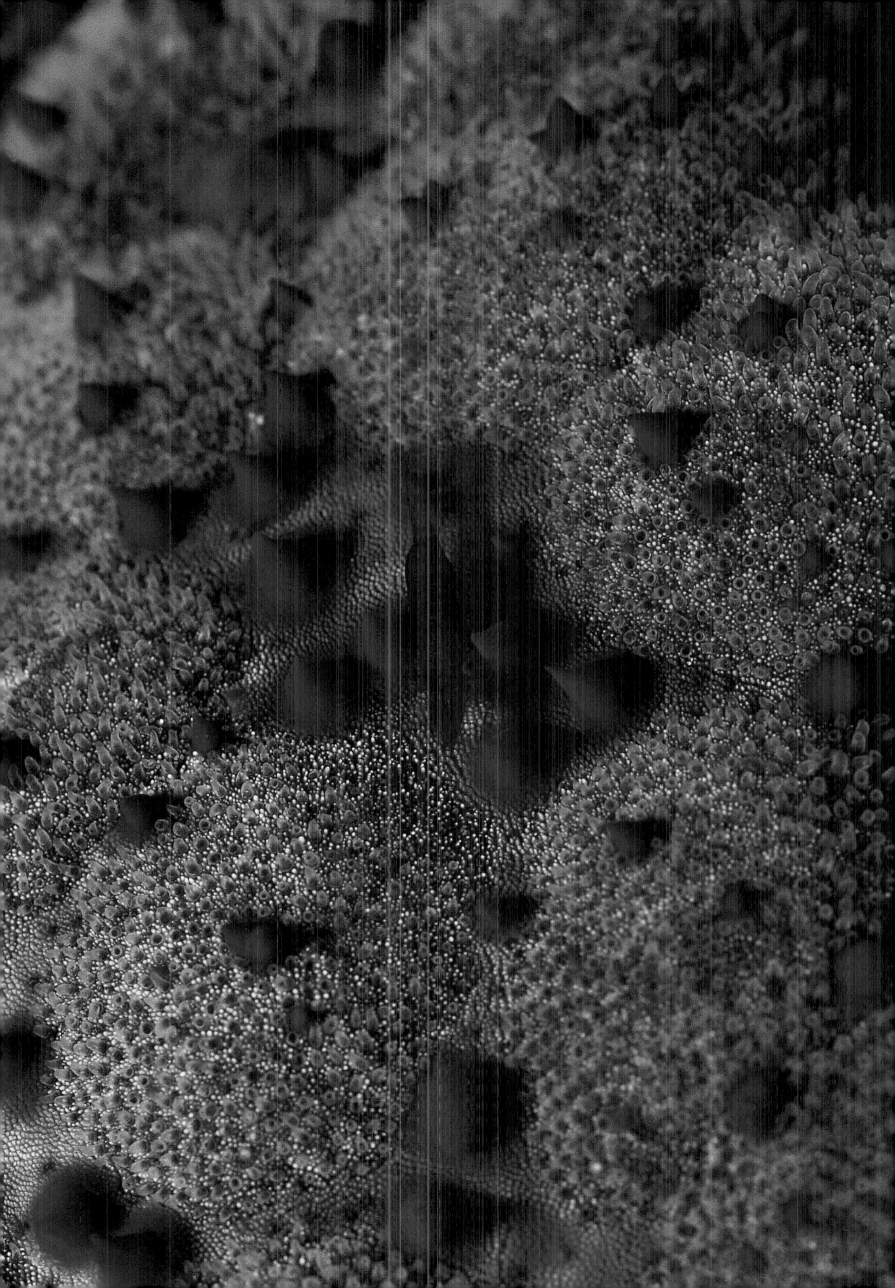

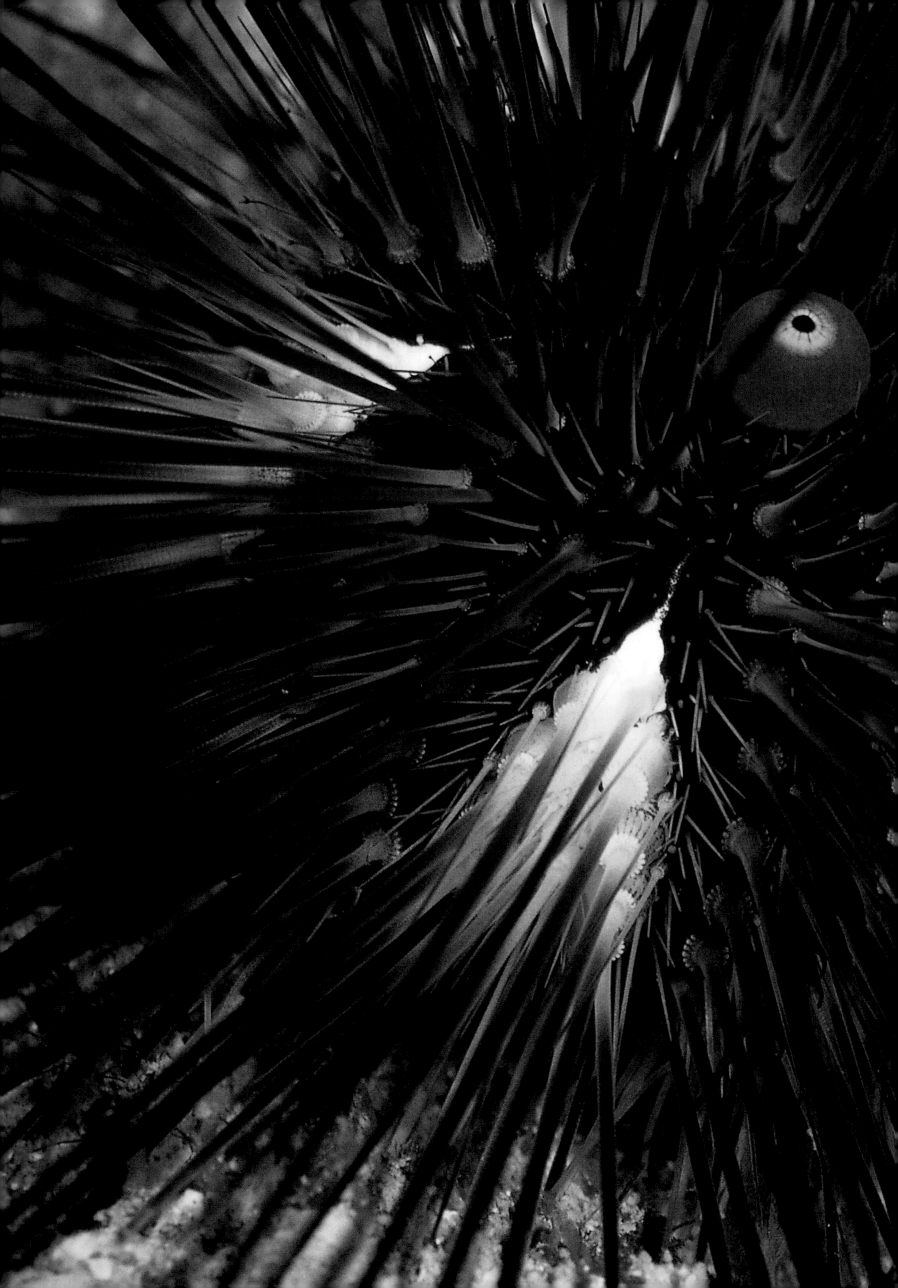

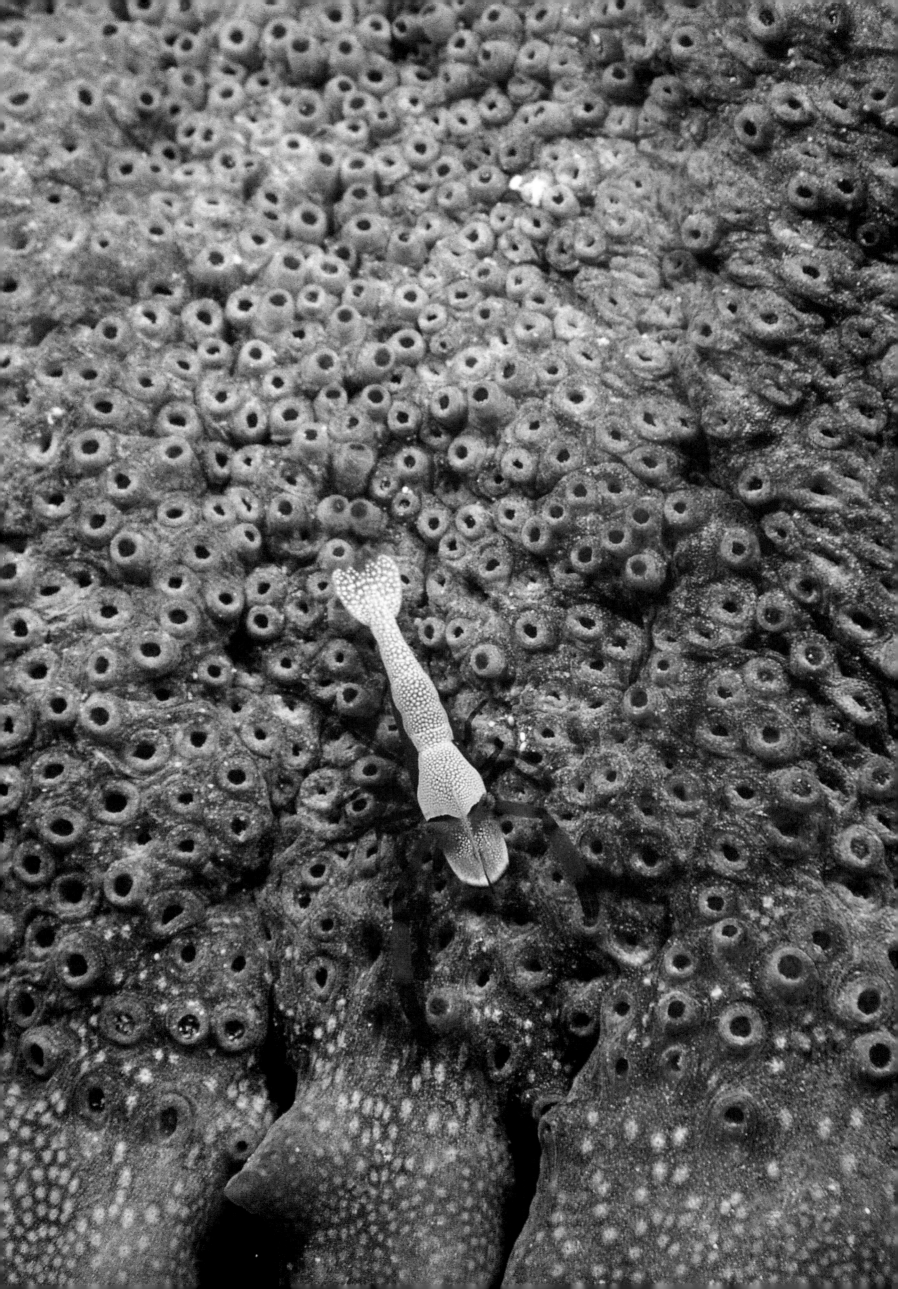

montipora, shapeless poritidae, mushroom, lettuce and brain coral. Her path crosses that of a troop of busy sergeant majors – these gregarious yellow and pale blue fish, with blackish-grey vertical bands, are also known as convict fish (*Abudefduf vaigensis*).

## THE TINY SHRIMP'S MESSAGE

As a swimmer, our diver is fascinated by the individual frenzy and overall harmony of the reef tenants below. With the ends of her fingers she strokes a fragile gorgonia and a crimson boring sponge, a vermilion wire coral (*Cirripathes anguina*) and a large green holothurian resembling some science fiction tank in a Fifties film.

Between the grey-green tentacles of an enormous carpet anemone (*Stichodactyla haddoni*), Sophie again finds the tiny species of shrimp (periclimenes) that was the subject of her first marine macrophotography. Delicate and transparent, heightened by ochre and beige eye-spots bordered in mauve; with long antennae it transmits a message of beauty to the world... Its Latin name? *Periclimenes brevicarpalis*. It is also known as 'peacock's tail'. Length: 3 centimetres (1.2 inches). For Sophie it is a souvenir dissolving into the present. A piece of nostalgia revived by a bagatelle of existence – an unimportant crustacean that takes on all the significance of life... Sophie visits the corals and gorgonias. She takes an interest in the sponges, the sea urchins, and sea-stars such as the precious blue starfish (*Lincksia laevigata*) and its orange relative (*Lincksia multiflora*) equipped with a long asymmetrical arm and thus called a 'sea comet'. The diver discovers some new little crustaceans and takes their picture. The imperial shrimp or emperor periclimenes shrimp (*Periclimenes imperator*) is as gracious as its peacock's tail cousin. It displays tints – pinks and beiges – and body

spots that vary with the individual; but the species always retains its orange, violet-tipped pincers: a social recognition signal; a flag... And here once again, amid the pale green vesicles of a bubble coral (*Plerogyra sinuosa*), the red-eyed periclimenes shrimp, its body as transparent as glass, is a symphony of rosy pinks and mauves. A crinoid shrimp (*Periclimenes tenuis*) waits in ambush in a feather star: with its yellow eyes, patterned red flanks and orange dorsal line, it closely imitates the sea-lily in which it finds bed and board.

The coral cleaner shrimp (*Stenopus hispidus*) 5 to 9 centimetres (2 to 3.5 inches) long, but provided with immoderately long antennae, is coloured in the typical red and white bands sometimes referred to as 'level-crossing gates'. This species lives

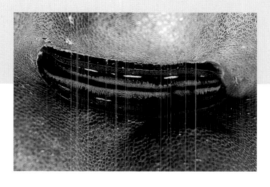

throughout the entire tropical ocean: it is a witness from the Jurassic era, a time when, stretching from China through the Mediterranean to the Caribbean, the Tethys sea separated the northern supercontinent Laurasia from Gondwanaland to the south... The banded cleaner shrimp brings the age of dinosaurs to the Maldives! At least collectively... Individually each one couldn't care less. In the range of marine ecological roles, this species has chosen the task of cleaning. It removes parasites from whatever presents itself, particularly from fish: because of this, it also been given the name of 'barber shrimp'. It places itself in a strategic position and waves its claws: a signal. The wrasse, lutjans, butterfly fish and angel fish know the code. They approach and, with mouth open and

*A delicacy and transparency, heightened by ochre and beige eye-spots...*

PRECEDING
DOUBLE-PAGE SPREAD
The black-spined diadem sea urchin (*Diadema setosum*) brandishes its arms. The little sphere is not a Cyclops eye, but an ordinary anal sac...

Length of spines: 20 cm (8 in)
Depth: 45 m (150 ft)
Site: Baros

OPPOSITE PAGE
An emperor periclimenes shrimp (*Periclimenes imperator*) melts into its background: the skin of a prickly red sea cucumber (*Thelenota ananas*).

Length: 2 cm (0.8 in)
Depth: 25 m (80 ft)
Site: Rihiveli

ABOVE
Tucked into the coral, a pectinid oyster (*Pedum spondylium = P. spondyloideum*) filters plankton.

Length: 5 cm (2 in)
Depth: 20 m (65 ft)
Site: Rangali

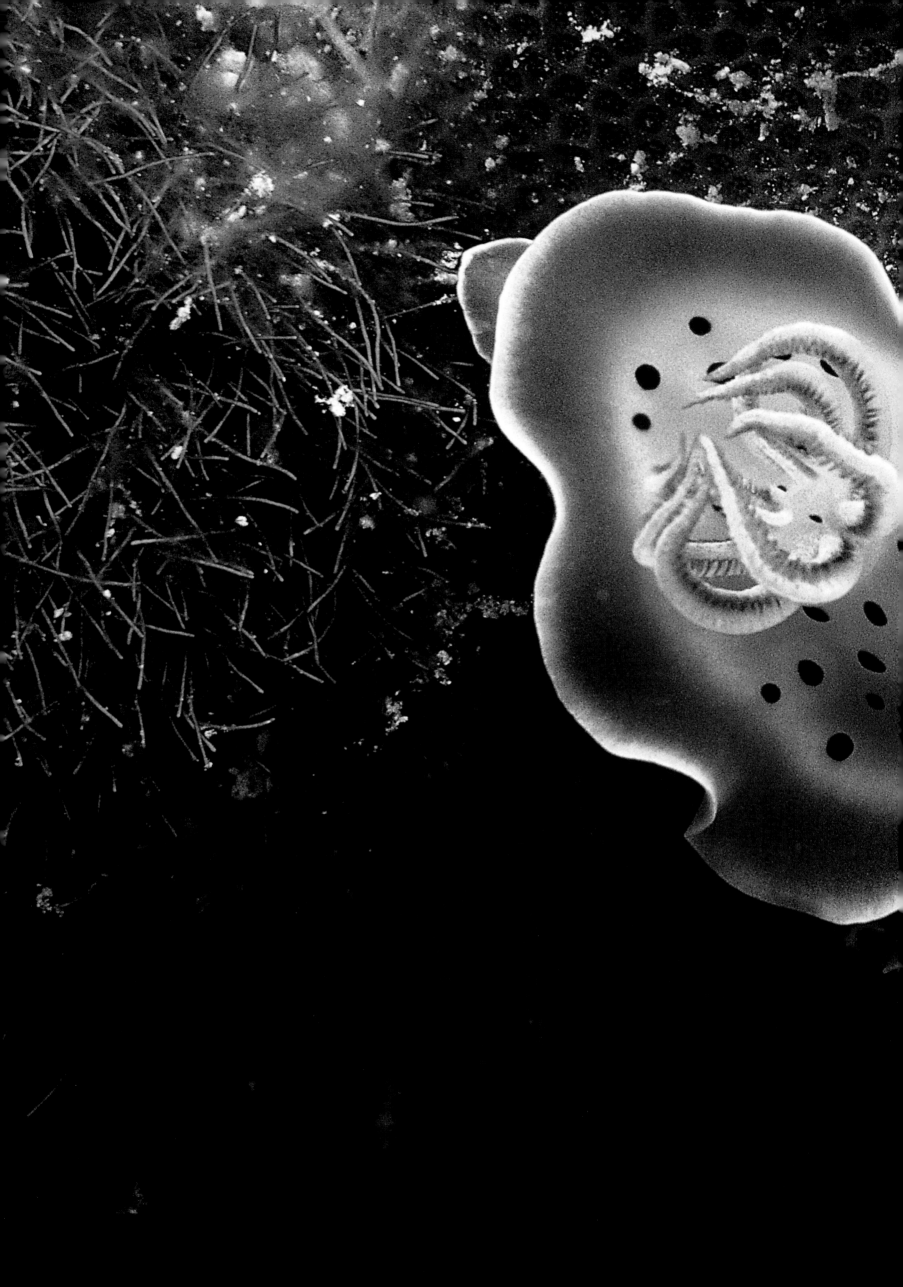

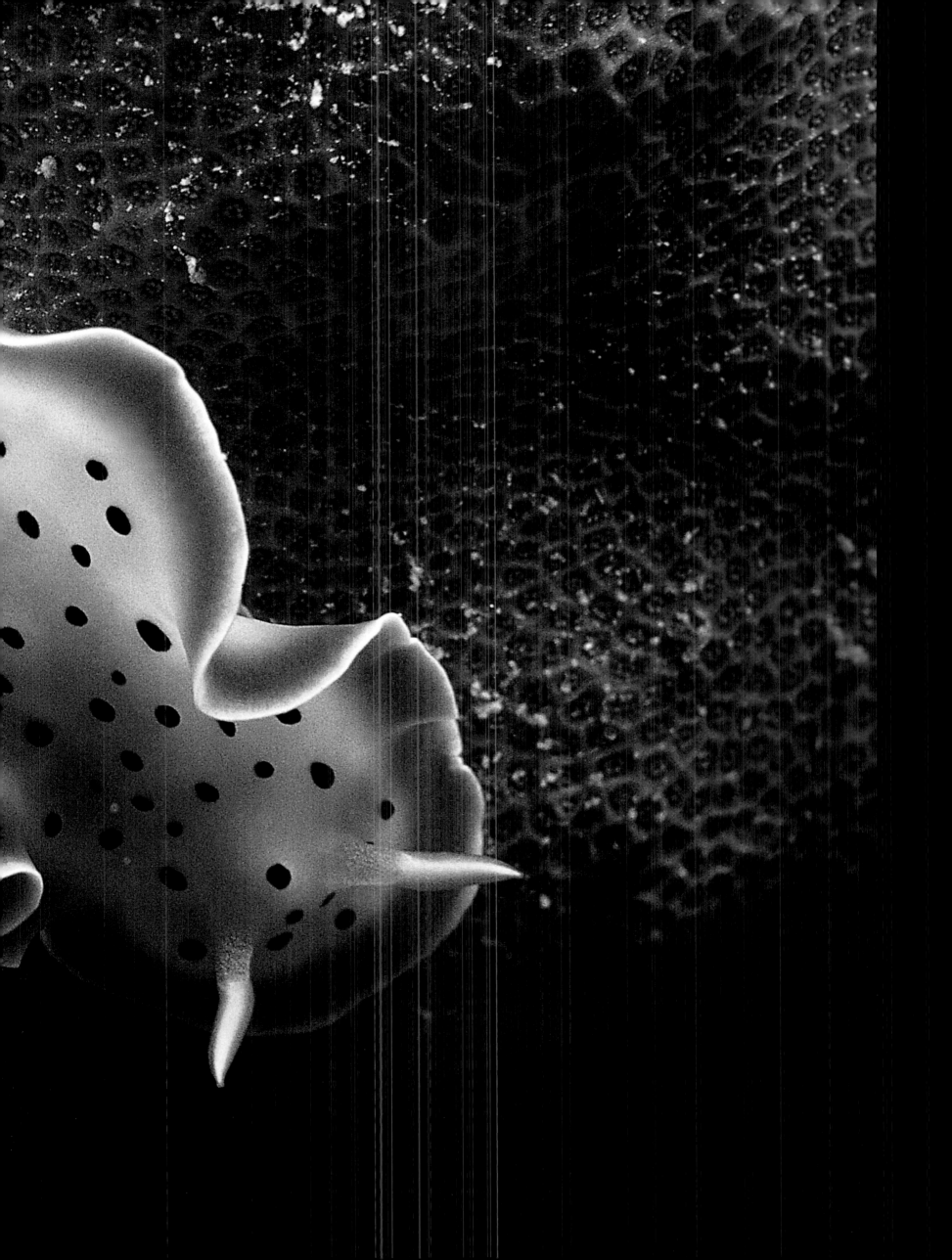

gills spread, they allow themselves to be cleaned by the crustacean. The shrimp is skilful and performs its task to the client's satisfaction. The service station is open!... Frequently this little business is carried on by a couple. The male is smaller than the female, but they share the work. This tranquil existence at the shop has a suggestion of provincial bourgeois life about it... The usual coral extravagance takes over. Sophie is familiar with its seductive ways. She appreciates its strange harmony, always at the limit of the grotesque, abiding by the laws of Beauty formulated by Edgar Allan Poe and Beaudelaire... Here monophoria, acrophoria, golf ball, mushroom and scroll corals establish their empire, reinforced by or competing with populations of

idly swing their tails. Blacktip reef sharks (*Carangoides bajad*) show their forked tails, their silver sides dappled by the sunshine. The scene is no less festive among the echinoderms. The sea-lilies resemble flowers with feather petals. Among these primitive animals, also known as 'crinoids', the sea-feathers are the most wonderful. Bennett's (*Oxycomanthus bennetti*) is ordinarily lemon yellow, but sometimes can be an inky black; it attaches itself to the substrate by means of organs known as 'cirri' and spreads a hundred feathery arms in the current.

Among the holothurians or sea cucumbers Sophie finds a pink *Holothuria edulis*: a sort of phallus 30 centimetres by 5 (12 inches by 2); much sought after, obviously as an aphrodisiac, in China and south-east

*The usual extravagance of coral takes over. The diving photographer is familiar with its seductive ways...*

PRECEDING
DOUBLE-PAGE SPREAD
The Maldive sea-slug *(Chromodoris tritos)* is also found in the Seychelles: its cloak is edged and spotted with mauve.

Length: 4 cm (1.6 in)
Depth: 35 m (115 ft)
Site: Wadoo

ABOVE, LEFT
Detail of the little white starry polyps of a building coral, or a madrepore or scleractinian coral.

Length: 8 cm (3 in)
Depth: 35 m (115 ft)
Site: Baros

ABOVE, RIGHT
A young clown fish takes refuge in the heart of a bubble tip anemone *(Entacmaea quadricolor)*.

Length of fish:
1.5 cm (0.6 in)
Depth: 20 m (65 ft)
Site: Rihiveli

OPPOSITE PAGE
The brightly coloured mantle of the giant clam *(Tridacna maxima)* has two 'breathing' holes (siphons), one to inhale (seen here closed), the other to exhale.

Siphon diameter:
8 cm (3 in)
Depth: 10 m (33 ft)
Site: Wadoo

gorgonias, alcyonarians, hydras, sea-pens, cerianths and anemones with sinuous tentacles. All the reefs look alike but each is different. Sophie planes over a corner abounding in molluscs. Here, more than just periclimenes and cleaner shrimps, there are also powerful Indian ocean lobsters (*Panilurus longipes*) with their brown-black diamond-studded Renaissance carapace, their antennae candy pink at the base. The mantis shrimps, squat lobsters, cowrie crabs and box crabs, ghost crabs and rounded crabs, all are cavaliers in armour. But these innocent combatants embrace no other cause than that of daily life, to the benefit of the earth and the universe. Sophie slips into a maze of fissures, corridors and unlikely arches. Three whitetip reef sharks

Asia, and consumed in such quantities as '*trepang*' that the species is in danger of extinction. The starfish continue to present their striped symmetry: the horned (or studded) star *Protoreaster*, the warty star, the red mailed star (*Fromia monilis*)...

But it is the sea urchins that will make this dive exciting for Sophie. She sets off in search of the pencil urchin, the lance urchin, the fire urchin and the poison or flower urchin (*Toxopneustes pileolus*), whose tiny defensive organs with pincers (pedicels) inject a venom that can kill a man. The diver's dream ends with the enormous spikes (sometimes more than 30 centimetres or 12 inches long) of the diadem sea urchin (*Diadema setosum*), where some tiny orange cardinal fish are hiding, fragile symbols of the reprieved Maldives.

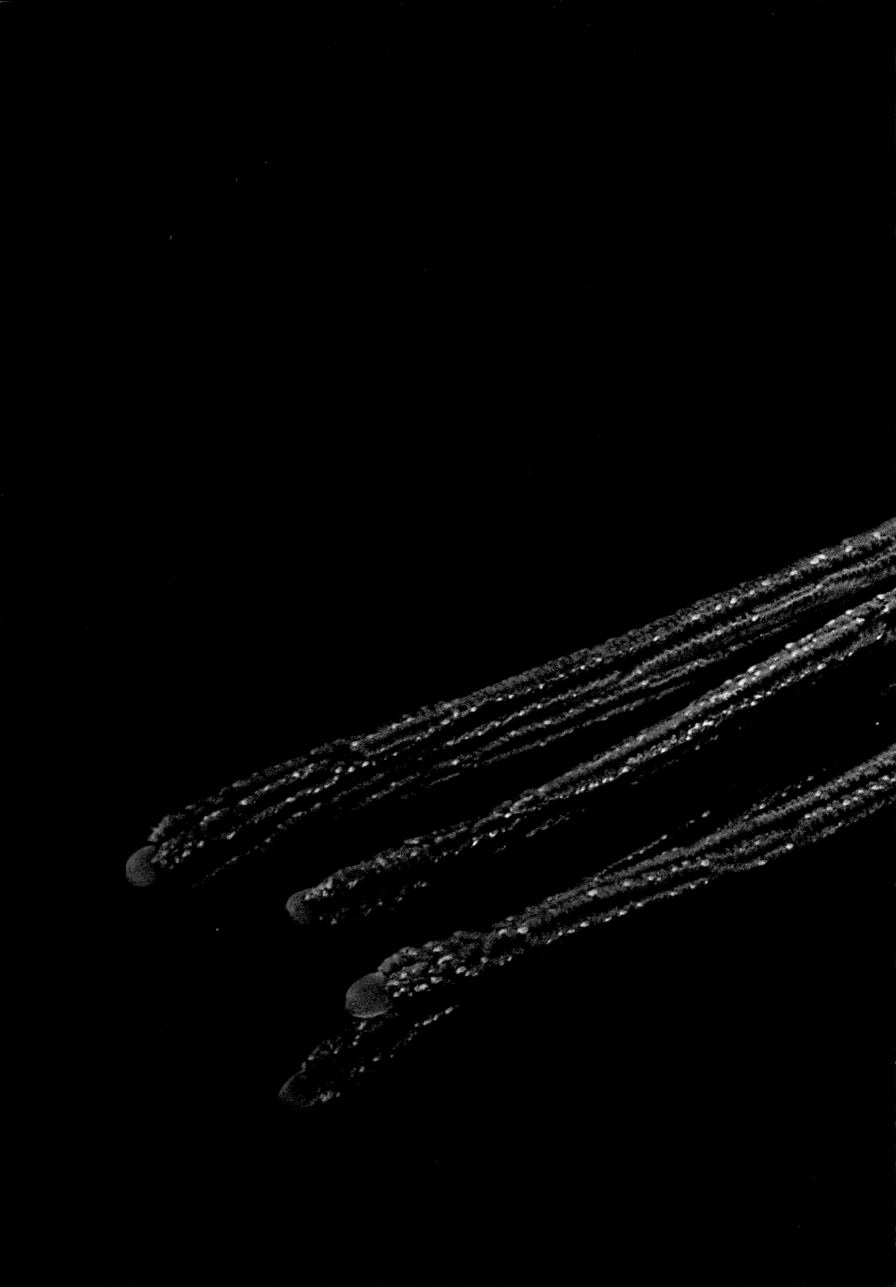

# MADAGASCAR

The Great Island.

*In this fragment of Gondwanaland,*

A mosaic of mysteries,

*lemurs leap and chameleons climb,*

above and below water.

*while under the sea fishes dart and*

*squid outdo their terrestrial counterparts.*

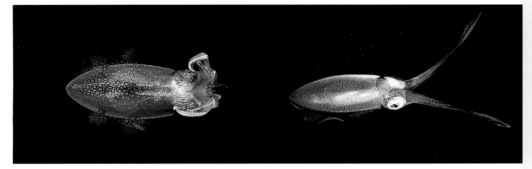

A strange and beautiful island. Larger than France. The terrain itself is known as the 'land of spirits', perhaps because of the lemurs that live there, seen by the Malagasy as the incarnation of their ancestors' souls. Under the sea, one forms a similar impression. Madagascar is a universe of the unknown, the indescribable, of strength of spirit and, if you believe in it, of the supernatural; at any rate of poetry. It is a world of the grotesque and the sublime, of life itself. Beneath a roof of coral, where the water seems dusky even at mid-day, pulses a shape, a spirit, a something. An alien, a being from elsewhere. Sophie swims towards this 'extraterrestrial'. Nothing excites her more than encounters that are the stuff of hallucination. The creature is translucent, suggesting a marble made of green Venetian glass. The substance of its body wall is jelly-like. Visible through this transparency is a twisting silvery digestive tube leading from the mouth. On its flanks bristle eight meridian lines composed of many tentacles, undulating like the teeth of some surreal comb. A comb. The animal is, in fact, what is known as a 'comb-jellyfish' a ctenarian, a ctenophore. In this case, a pleurobrachian, the sea gooseberry. A wanderer, a vagrant among the currents. One of the tiny plankton people.

Sophie approaches the creature. The tentacles of this coelenterate are furnished with sticky cells, not the stinging cells (cnidoblasts) found in the cnidarian coelenterates such as the jellyfish, anemones and corals. The sea gooseberry lets itself drift in the liquid element, dancing in an eddy. One might say that this movement epitomises the whirlwind of evolution since the Precambrian age. The comb-jellyfish are the survivors of the procession of soft creatures that have haunted the oceans for 700 million years. For her 'Madagascar' assignment, Sophie has chosen to visit the Nosy-Bé islands and, in this archipelago, Nosy-Komba and some other sites such as Ankarea, Tsarabajina...

Nosy-Komba is an ecological prodigy. It is a volcanic block draped in virgin forest, where are seen sunset moths, chameleons with the bearing of dinosaurs, and red-brown, crafty-eyed macauco lemurs with tufted ears and bushy tails.

## THE LEMUR AND THE MORAY

The diver has left the butterflies to their nectar, the reptiles to hunting them with their tongues and the primates to their immoderate love for papayas. She has passed through the line of coconut palms, walked across the beach and reached the reef by swimming over a sandy seabed. On the bottom she has found various different snails (coneshells, queen and spider conch), starfish and a peacock flounder (*Bothus mancus*) so well camouflaged that she would never have spied it without her

PRECEDING
DOUBLE-PAGE SPREAD
The thyanostoma jellyfish *(Thyanostoma lorifera)* pulsing on its mysterious way, drifting aimlessly in the ocean current. Testimony to the primitive ages of life...

Length: 40 cm (16 in)
Depth: 2 m (7 ft)
Site: Nosy-Bé

ABOVE AND
OPPOSITE PAGE
Three poses of the broad-finned squid *(Sepioteuthis lessioniana)*. This cephalopod mollusc, whose colours change more than those of the chameleon, swims by jet propulsion and captures its prey with tentacles that could be described as expressive...

Length: 25 cm (10 in)
Depth: 5 m (16 ft)
Site: Ankarea

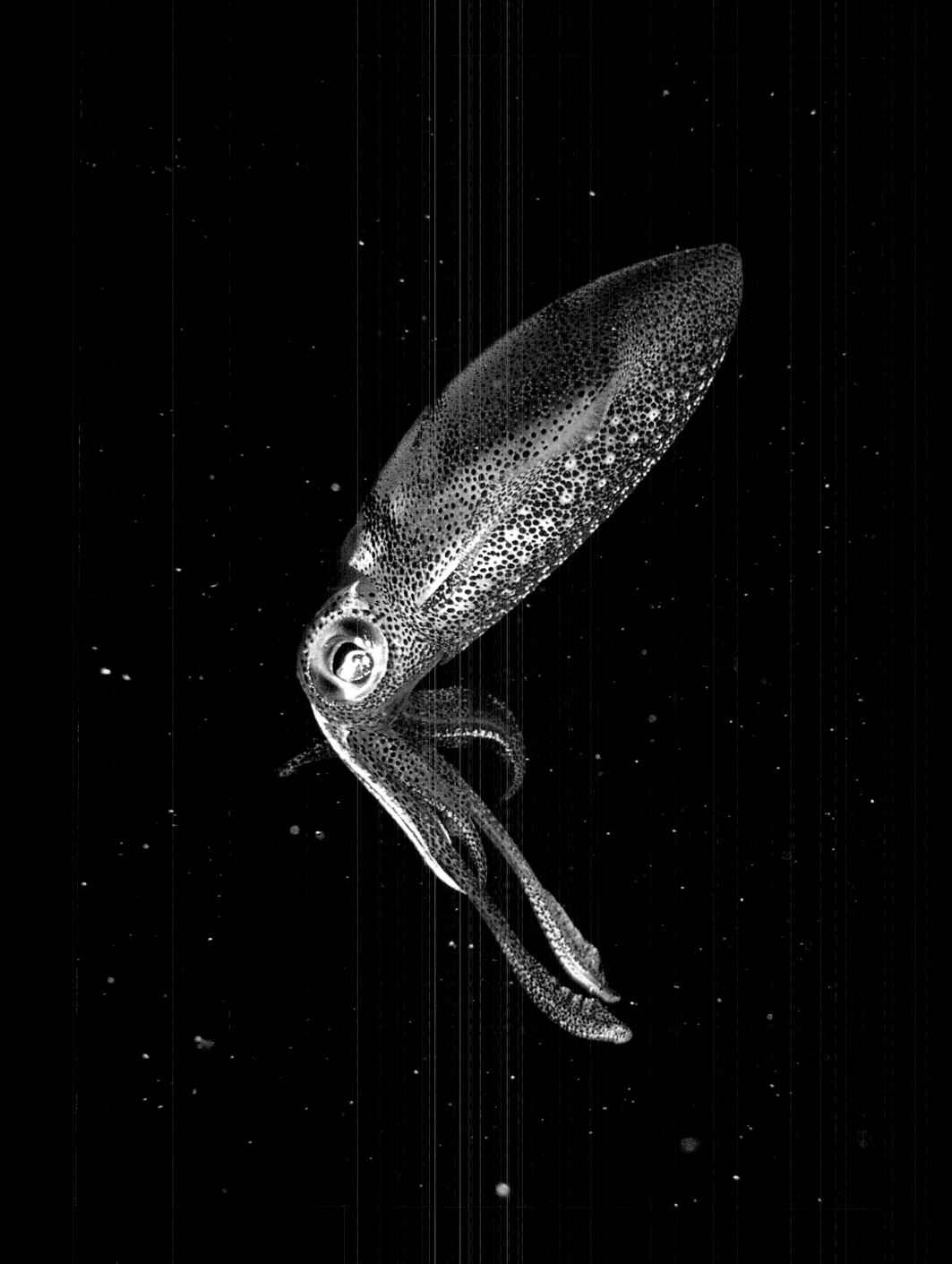

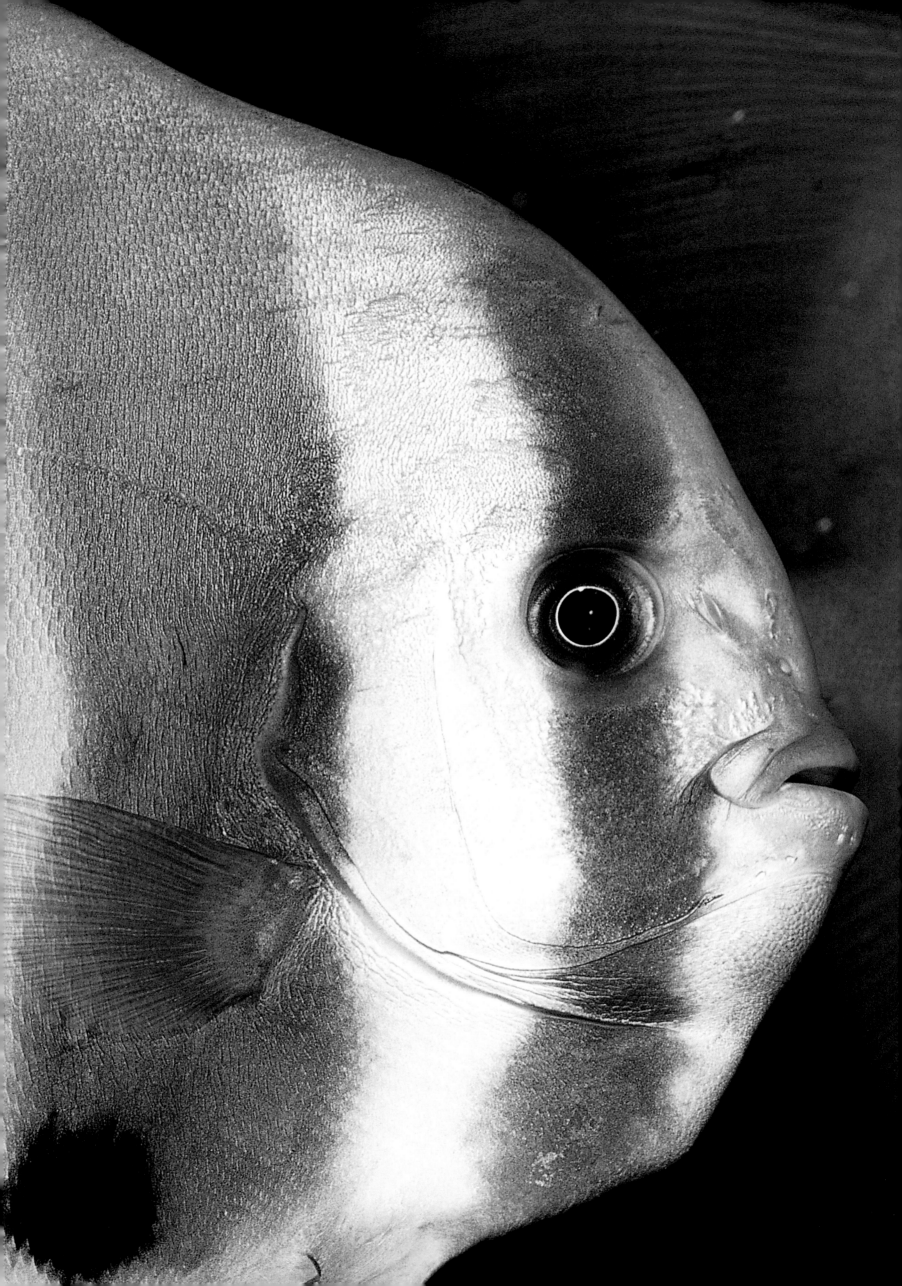

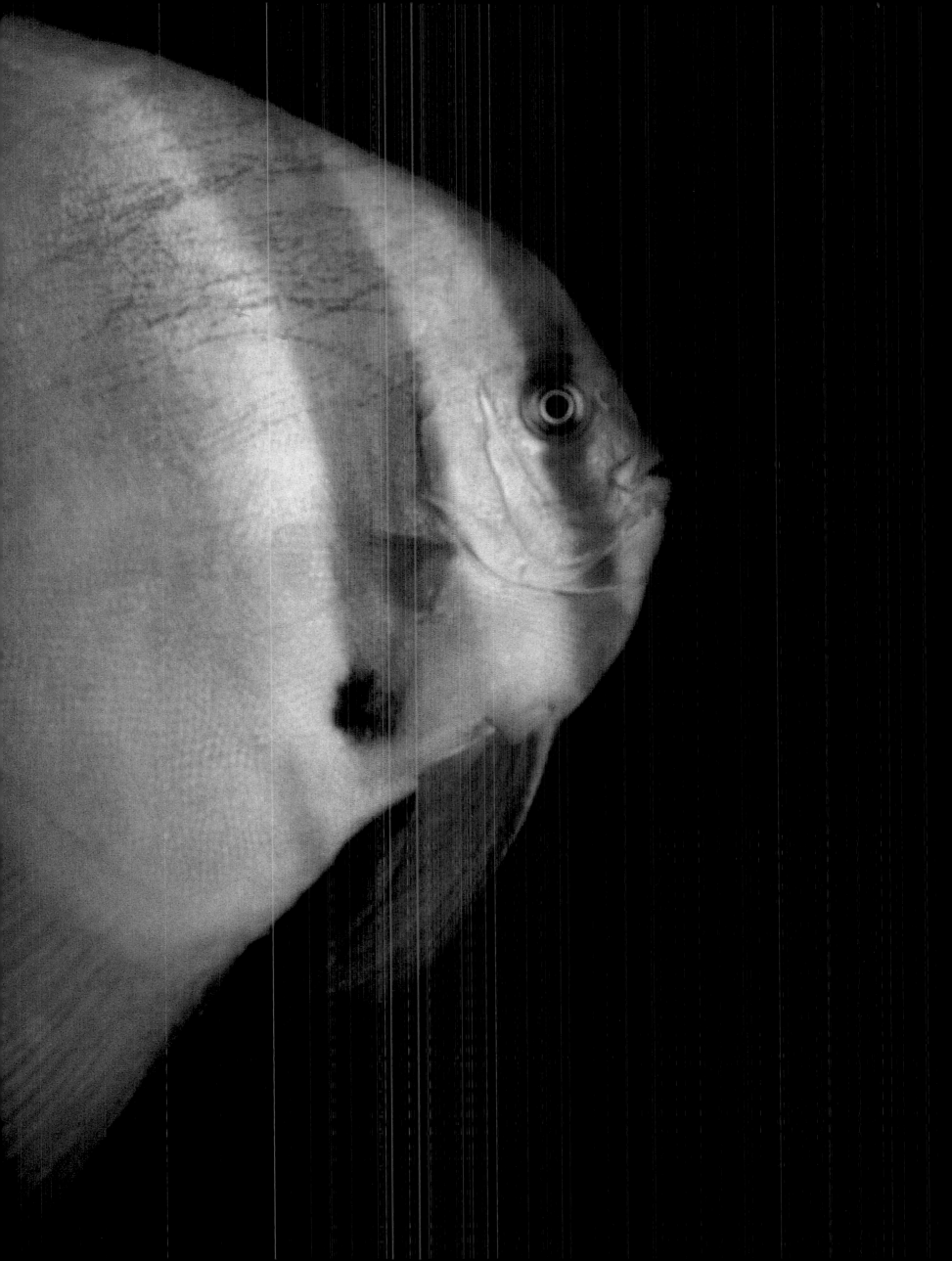

marine naturalist's eye. The diver greeted the sea gooseberry while floating beyond the drop-off. Now she sinks down between the madrepores, the fire corals, the gorgonia fans and the barrel sponges. Not far from a bulky elephant's ear anemone (*Amplexidiscus fenestrafer*), a blood-coloured sea-slug waves the flounces of its flattened, crinkled body that is like a flamenco dress: it is no surprise that this mollusc is called 'little Spanish dancer'!

## THE CHAMELEONS OF THE SEA

Sophie glides over a ledge planted with orange and pink alcyonaria, from which gape some tropical oysters and a fan shell. A ghost crab scuttles along on a bed of turquoise and jade sea squirts. She passes through a coral arch and discovers a cave whose walls are garlanded with golden *Tubastrea* and gnarled red *Melithea* gorgonia coral. She slips inside. Two clefts at the side harbour morays, one marbled (*Gymnothorax undulatus*), the other (*Siderea grisea*) grey with violet dots on its snout.

A dozen ghosts are dancing in the cave's shadows. Their oblong bodies, tentacles and triangular fins are speckled, hatched, striped and inlaid in colours that change with the speed of light. Violets replace reds, then give way to browns and greens, blues and coppers. These creatures setting off fireworks are cuttlefish (*Sepioteuthis lessioniana*). Their chromatic variations are produced a thousand times as fast as those of the chameleons on dry land.

Cuttlefish and the other cephalopods (squid and octopus) are the chameleons of the sea. Some of their skin cells, called 'chromatophores', contain black, red or yellow pigments (melanine, erythrine or xanthine). These substances are in the form of granules that are displayed or retracted (the colour disappears). Other 'physical' effects (the blues,

violets and rainbow colours) result from a subtle interplay of light (reflection, refraction, prism effects) seen through guanine crystals. Examining the crannies, Sophie discovers a pharaoh cuttlefish (*Sepia pharaonis*); a squid bristling with ringlets that give it the appearance of a toy poodle, and a blue-ringed octopus (*Octopus cyaneus*), its great, gold eyes traversed by a horizontal pupil. Colour here in plenty, as the young female takes off over the coral. The tropical fish are splashed with more colour than children's paintings. Among the most psychedelic are the butterfly fish – for example the coachman butterfly fish (*Chaetodon auriga*), its body white in front and bright yellow behind, barred with oblique stripes, or this Meyer's butterfly fish (*Chaetodon meyeri*), with its

wide, curved black markings on a grey and gold background, reminiscent of a Hebraic script. The butterfly fish show as many tints, stripes, bands, streaks, lozenges and ovals as the combined works of Picasso, Braque and Matisse, and as many splendours as can be seen on the wings of the Madagascar sunset moth. Happiness under the sea, certainly... The fish dart about as in a delightful dream among the gorgonias and madrepores. Amid this explosion of colour it is difficult to award the prize for beauty. Should it go to the longnose hawk fish (*Oxycirrhites typus*), with its net of red on a rosy-pink ground? Or to the lemon and green damsel fish (*Pomacentrus sulfureus*)? To the royal angel fish or to its cousin the emperor angel fish?

*The fish dart about as in a delightful dream among the gorgonias.*

*It would be difficult not to see signals, codes, exchanges in the interplay of colours and light.*

To the three-spotted (*Apolemichthys trimaculatus*), a sunny yellow with blue make-up on its lips and forehead? Or to the striped sweetlip, red barber fish, diamond wrasse, the scarlet soldier fish, the green parrot fish, the trigger fish, the surgeon fish, the clown fish? Eventually, everyone may well agree on the flying scorpaenid, the ocean lion fish, striped in brick-red and white and bristling with fins divided into eagle feathers.

### MYSTERIES TO BE SOLVED

Why are tropical fish such works of art? What are all these nuances of colour for? Sophie loved to smile and say: 'We should stop thinking of ourselves as the navel of the earth.

Publicity. Want ads and classifieds. Sometimes the statement is hostile: 'wicked teeth!' or 'poisonous spines!'
'When diving' said Sophie, 'I always have the impression that I am surrounded by messages. If it is not exactly a department store at Christmas, or a welcome page on the Internet, there is still a similarity... The problem for us humans is that we have no translations of these bizarre signs in our usual language.'
Konrad Lorenz's intuition seems to be correct. If the scaly text repeatedly reads 'Watch out for yourself!' one suspects that in other circumstances it whispers: 'I love you!' In the multicoloured jungle of the reef, a good number of the spots, marks, niellos, rosettes, runes

**PRECEDING DOUBLE-PAGE SPREAD**
The eye and mouth of the diamond wrasse (*Cephalopholis miniata*). With its baroque flanks and exaggerated tones, this species incorporates the colour of the reef.

Length: 40 cm (16 in)
Depth: 50 m (165 ft)
Site: Nosy-Bé

**ABOVE**
The manta ray (*Manta birostris*) flies! It is also called 'sea devil' on account of the two 'horns' that direct plankton into its mouth.

Length: up to 6 m (20 ft)
Depth: 5 m (16 ft)
Site: Tsarabajina

**OPPOSITE PAGE**
Against the light, a shoal of fish forms a carousel. Gatherings like this serve for mutual defence as well as amorous frenzies.

Length of a fish: 25 cm (10 in)
Depth: 8 m (26 ft)
Site: Nosy-Komba

The reefs ignore us. The animals that live there do not don their luxurious clothes for the diver's benefit: they lit up the sea before man arrived.'
We assume that for them, to shine out means to 'say something'. To convey a message. To put up a poster. It would be difficult not to see signals, codes, exchanges in the interplay of colours and light. A conversation. The first person to describe the colour of fishes as a language was none other than the founder of behavioural science, Konrad Lorenz. To the father of ethology, the motifs that enhance the creatures of the reef are *plakatfarben* 'poster colours'. These lines, spots and fantasies are not gratuitous. They are the hoardings on our streets. They are the posters that call the customer to the show or to the sale. Advertisements.

and ideograms are impassioned declarations. Say it with blacks, reds, greens, blues. Rimbaud's vowels... In many species the recognition of sex is a visual affair. Of form, colour and body painting...
On the coral reef, as among the trees of the tropical forest, smell plays a lesser role than vision. In order to beget descendants, it is important to be admired. The parrot and the angel fish resort to the same strategy: they look splendid in order to perpetuate their genes. They make themselves lovable or sublime to enjoy a short moment of ecstasy.
Who could blame them? Certainly not humans, who have promoted clothing to the level of an industry and make-up to the height of an art.

# MALAYSIA

Crackling. Sparks.

*The sweltering heat of south-east Asia creates a*

Neon. Explosions.

*magic spell... The seas are a vision in silver*

Bengal lights. Fireworks...

*blue, scattered with amorous fish and*

*invertebrates created by a conjuror.*

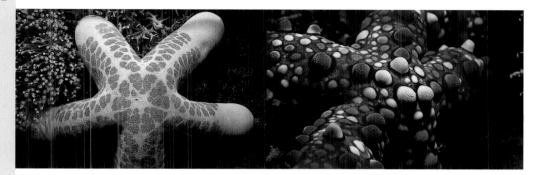

Here the animals of the sea do not shrink from excess. Sponges, corals, worms, molluscs, crustaceans, echinoderms and fish take on the most exuberant as well as the most subtle shapes and colours. Kitsch or charming, always fantastic. Even the great Charles Darwin, on his tour of the world between 1831 and 1836 in the *Beagle*, cast aside his scholarly reserve and loosened his tongue, mentioning diamonds, gold lamé, silver niello, rubies and sapphires... These comparisons hold good. It is impossible not to imagine that (as in the cartoon films) treasure lies below the surface, guarded by giant octopus or cruel shark... In this Ali Baba's cave, carbuncle and topaz fish zigzag between coral pilasters and wander over the brain coral and bubble coral, sparkle over fan shells and volutes, nibble among the alcyonarians, or sleep in gorgonia lace.

What a strange theatre! Parrot fish and clown fish. Damsel fish and butterfly fish. Wrasse and surgeon fish, flying scorpion fish and angel fish. With barber fish, unicorn fish, copper-striped lutjans and red diamond groupers studded with precious stones. Bronze jacks and steely-blue barracudas. Cow fish and goat fish. Inflated puffer fish, bristling porcupine fish and humped napoleon fish. Poetry in names, appearances and colours.

The coral's tenants are magicians. Whoever encounters their splendour forms the impression of beginning a journey of initiation befitting a Mexican shaman, or an experiment with mescaline performed by Henri Michaux. You set out logically enough, then dissolve into a dream, searching in vain for the words to convey this swarming plenty. 'If you want to try out the submarine world,' said Sophie, 'forget everything you have ever seen, read or heard about it: nothing can prepare you for the wealth of the reef. I have swum over so many reefs. I have never found what I was looking for, always something much better!'

## A JOURNEY OF INITIATION

Sophie has reached south-east Asia. Malaysia. . She dives in several places on the peninsula, and in the straits of Malacca. Too many tankers and cargo ships; traces of oil and pollution: in short, 'civilisation'. Nonetheless, below the surface the variety of life persists: proof that nature reacts and resists. If we give it time, it will repair itself better and more quickly than the ecologists think. Encouraging news.

On the seabed off Langkawi island, in the north of the country near the Thai border (Phuket is only 200 kilometres or 125 miles away), submerged landscapes of rock, moulded and carved by erosion, as rounded and curvaceous as the female form — after Rubens or Botero — harbour a remarkable fauna. Sophie is equipped and swimming on the surface.

**PRECEDING DOUBLE-PAGE SPREAD**
A whole community on a gorgonia branch! Space is in short supply under the sea. A simple gorgonia branch hosts little sponges, colonies of sea squirts and fine sea-feathers.

Length: 20 cm (8 in)
Depth: 30 m (100 ft)
Site: Banggi

**OPPOSITE PAGE**
A sea-snake brittle-star or ophiurid (perhaps of the genus *Ophiolepis*) twists its long ringed arms in a jumble of coral tube flowers.

Length: 10 cm (4 in)
Depth: 35 m (115 ft)
Site: Langkawi

**ABOVE, LEFT**
The granular cushion star *(Choriaster granulatus)* is recognised by its plump arms.

Length: 15 cm (6 in)
Depth: 25 m (80 ft)
Site: Sarawak

**ABOVE, RIGHT**
The knobbed starfish *(Echinaster callosus)* is studded with elegant coloured bumps.

Length: 20 cm (8 in)
Depth: 10 m (33 ft)
Site: Banggi

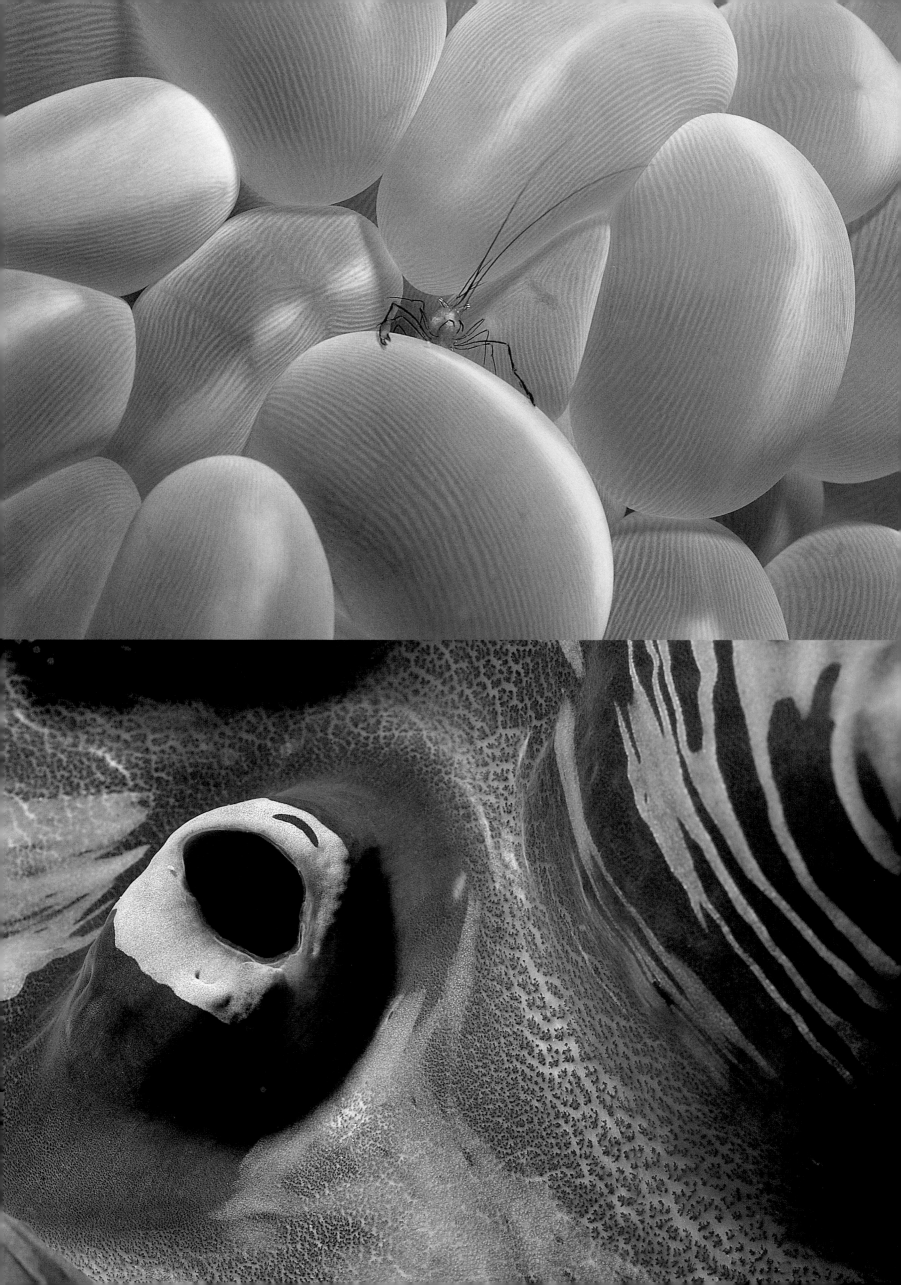

A baby green turtle a few days old paddles in front of her. This delicate little body with an uncertain future seems animated with incredible energy. Its thin shell is vulnerable. Its tiny horned beak is only capable of snapping up plankton larvae. But the new arrival works its flippers and propels itself along through the liquid, innocently enjoying the slice of time it is offered. A short while ago it was in its egg, in the sand somewhere on the beach. It dragged itself out and, escaping the herring gulls and frigate birds that plunder every hatch, ran for the sea and entered the foam. Now here it is, in its element: but many dangers lie in wait: sharks, jacks and bonito would consider it a mere mouthful. With good fortune this child of hope will become an adult turtle impregnable in its armour, living for more than a hundred years, into the twenty-second century. Sophie sinks towards the depths in water teeming with plankton, larvae, swimming shrimps, and the little sea wasp jellyfish whose sting is best avoided: there is a risk of fatal allergic shock. A strange shape twists before her, upright in the water, like a rope raised up by a circus magician. It is a democratic salp... Not a single animal but a colony of squarish creatures, each one stuck to its fellow, in a long chain provided with a gleaming, transparent digestive tube.

### THE SARDINES AND THE ZEBRA SHARK

These gelatinous beings could pass for primitive jellyfish. But salps belong to the tunicate group: they are protochordates. In the embryonic stage at least, they possess a dorsal chord (from which comes the term 'chordate'), that is, the start of a nervous system that was the making of the ancestors of the vertebrates. They began our lineage. They are among our ancestors. In the Ordovician age, some 450 million years ago, they gave rise to a class of fishes, of which one

branch (the bony fishes with lobed fins, represented by the coelacanth) gave birth to the amphibians (360 million years ago), to reptiles, birds and mammals. And us, us, finally us!

Sophie lets herself drift in the warm, reassuring, maternal sea, between its breasts, belly and buttocks of monumental granite, on which clouds of thousands of elvers gather, disperse, explode and regroup.

These shoals are fascinating. They react in unison, as a single organism; with perfect co-ordination. This harmony is made possible by the sense of distant touch that fish possess. Their lateral line contains thousands of pressure receptors. Thanks to these ultra-sensitive nerve endings, each fish in the shoal

perceives the vibrations around it, and adapts itself almost instantly to the posture of its close neighbours. The shoal formation is also useful when finding food and sexual partners, as well as for confusing predators. At least for a moment they may be bewildered by a teeming mass of potential prey.

A zebra shark (*Stegostoma fasciatum*) emerges from between two thighs of rock. More than 2.5 metres (8 feet) long, it has a streamlined body, an asymmetrical (heterocercal) tail and a few longitudinal fins. Its grey-yellow skin with blotches of black has earned it its other name — 'leopard shark'. It has zebra stripes when young.

Sophie pursues the prowler for a moment as it follows a corridor bristling with feathery hydras and

*On the seabed, sunken landscapes of carved rocks, rounded like the female form...*

**PRECEDING DOUBLE-PAGE SPREAD**
This chromodoris sea-slug (*Chromodoris sp.*) awaits a species name. Like the others it is hermaphroditic: mutual fertilisation!

Length: 5 cm (2 in)
Depth: 20 m (65 ft)
Site: Banggi

**OPPOSITE PAGE, ABOVE**
The bubble coral shrimp (*Vir philippinensis,* chooses its home in the strange madrepore (*Plerogyra sinuosa*) that gives it its common name.

Length: 2 cm (0.8 in)
Depth: 8 m (26 ft)
Site: Langkawi

**OPPOSITE PAGE, BELOW**
The exit siphon of the giant clam (*Tridacna maxima*) emits reproductive material (eggs and sperm) as well as ejecting waste from its great shell.

Siphon diameter: 8 cm (3 in)
Depth: 5 m (16 ft)
Site: Sarawak

**ABOVE**
A nembroth sea-slug (*Nembrotha kubaryana*) is decorated with sublime but highly variable colours.

Length: 6 cm (2.4 in)
Depth: 20 m (65 ft)
Site: Langkawi

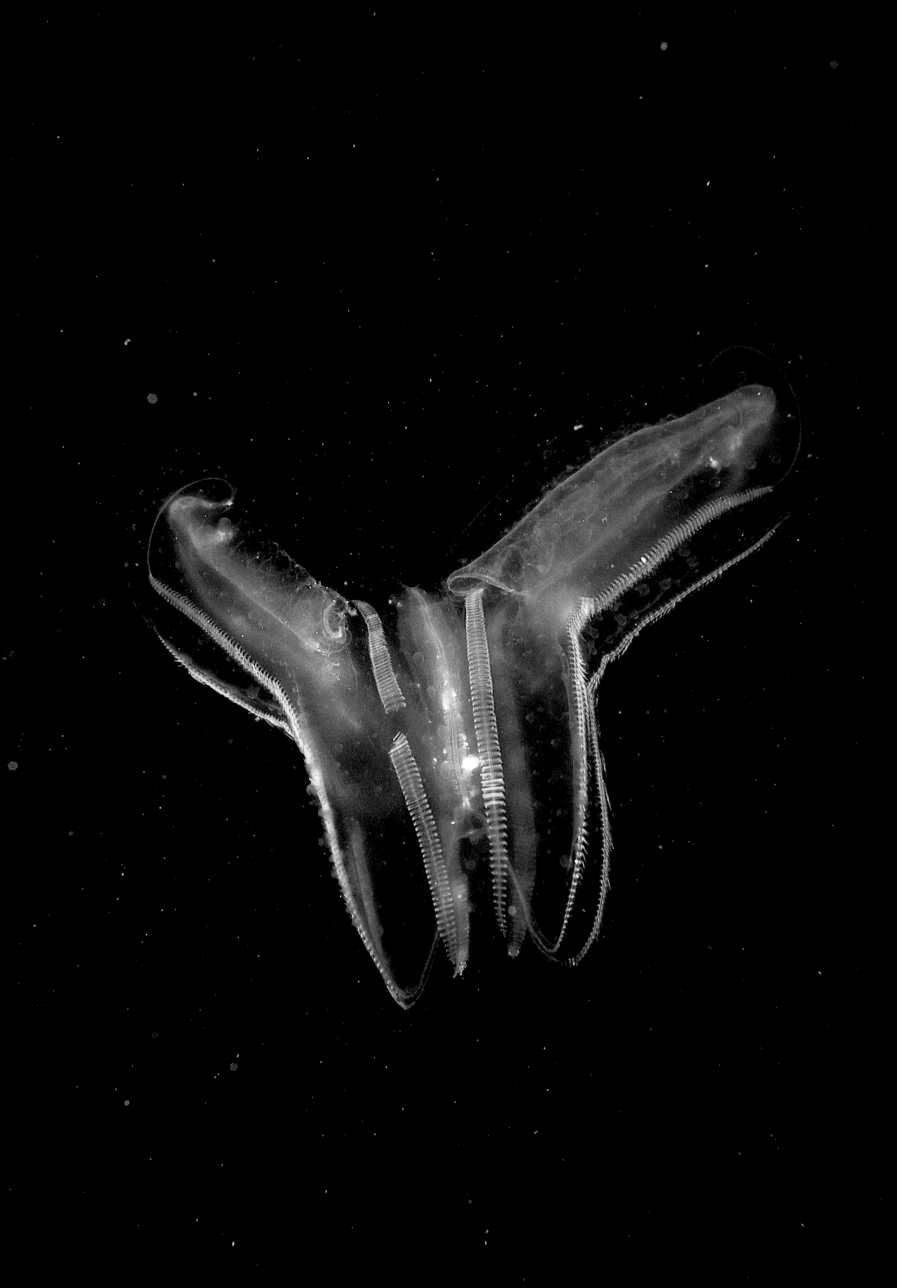

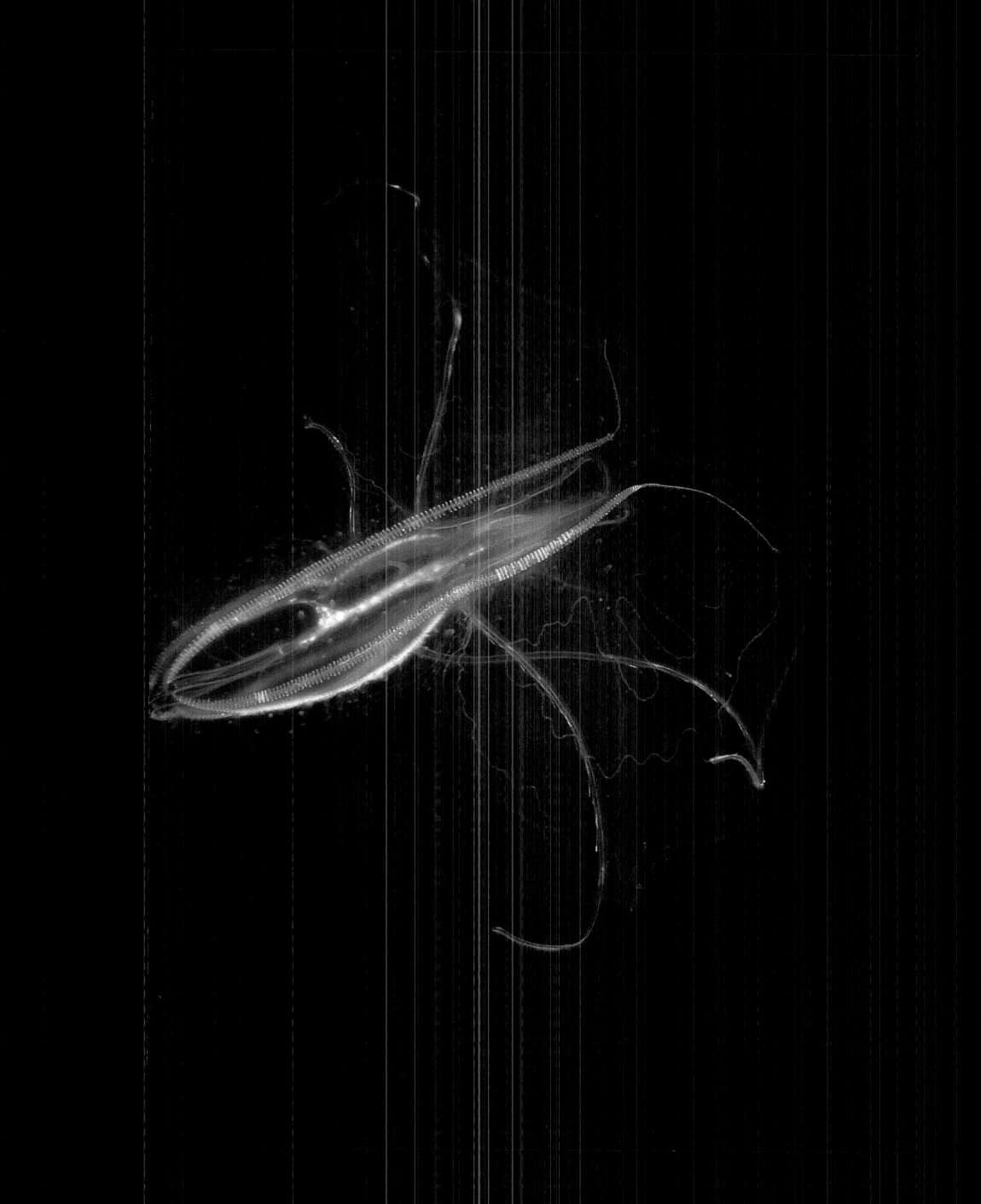

white wire gorgonias, set off by golden tube coral. In her eyes, the shark represents perfection. One of the most beautiful creatures in a not ungenerous ocean. The shark sprints for the blue-black of a steeply sloping canyon and reaches the depths. It is impossible to follow… Further on, in a cliff, Sophie finds a heart-shaped opening the size of a plate. This gap gives onto a cave carpeted with bright red sponges. A scarlet heart — for a love story?

Some reef squid (*Sepioteuthis lessioniana*) dance before the door. One female and several males, one outshining the next, moiré patterns, zigzags, electric punctuation…

The free-for-all is confusing but it seems that male and female youngsters embrace: ten tentacles a side for a passionate super kiss.

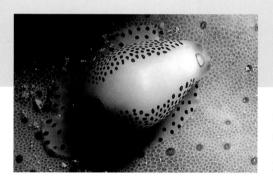

Using a specially adapted arm ('hectocotyl'), the male takes a packet of sperm (a spermatophore) from its genital opening and, with a quiver of ecstasy, inserts it into the palleal cavity of the female just at the lip of the oviduct…

## WONDERLAND

Fertilisation over, the female will lay her eggs in this little heart-shaped house. She will stick the eggs to the ceiling in necklaces of hundreds of white eggs. Without eating, she will oxygenate and defend them until they hatch. She will die, exhausted, bereft of all energy, on the day her babies extricate themselves from the eggs, tiny ten-armed comets lost in the infinite liquid of the lower sky.

In Malaysia, Sophie does not content herself with exploring the coast of the peninsula and the straits of Malacca. She travels on, to the island parts of the country — the provinces of Sarawak and Sabah on the island of Borneo.

At the very tip of Sabah, almost at the Philippines, she visits Banggi island. A distant spot beyond a jungle, perfect and improbable like the beginning of a blue dream. A 'wonderland' like that of Alice, where the heroine, a little girl once more, meets a spotted rabbit fish (*Siganus stellatus*) and follows a magician hurrying through the maze of the reef, who knows why and to where. She stops at the house of the 'mad hatter' — shall we say, the sea-slug (*Nembrotha nigerima*), a violet-black with green spots. A surgeon fish threatens to make an incision with his sharpened scalpel — perhaps the gold-tailed blue surgeon (*Paracanthurus hepatus*). Not to mention the sarcastic remarks from the 'king and queen of diamonds', in the person of (*Scorpaenopsis diabolus*) the poisonous, marvellously camouflaged devil scorpion fish. And the great pineapple sea squirt, a reddish-brown, warty sea cucumber sometimes as long as 80 centimetres (30 inches), that normally moves slowly on the bottom, but when provoked sits up and spits out long white sticky threads, a textile that is surreal to the point of incomprehension.

The universe below scarcely lends itself to scientific investigation or human philosophy. It is still unfamiliar to us, different in every way and often hostile. It reminds us that if we have already been aquatic creatures, one day we chose terra firma and the atmosphere. We were amphibians, then reptiles, before becoming mammals. We have reached a new world and future. But perhaps we have lost much: all kinds of feelings and pleasures. A kingdom we shall never truly reconquer. A three-dimensional, weightless world. Our liquid element, as it were!

*A distant place, perfect and improbable like the beginning of a blue dream.*

PRECEDING DOUBLE-PAGE SPREAD, LEFT
The funnel-shaped bolinopsis ctenophore *(Bolinopsis infundibulum)* forms a part of the plankton: prey is captured with its two large buccal lobes.

Length: 12 cm (5 in)
Depth: 2 m (7 ft)
Site: Langkawi

PRECEDING DOUBLE-PAGE SPREAD, RIGHT
This other ctenophore, or comb-jelly (species unknown) swims in the current by means of buccal lobes deployed like sails.

Length: 6 cm (2.4 in)
Depth: 3 m (10 ft)
Site: Sarawak

ABOVE
The white shell of the warty eggshell cowrie *(Calpurnus verrucosus)* is ochre or orange at the ends.

Length: 3 cm (1.2 in)
Depth: 25 m (80 ft)
Site: Banggi

OPPOSITE PAGE
The broad-armed squid *(Sepia latimanus)* hunts lobsters or fish. This one is pale with emotion on meeting the photographer.

Length 30 cm (12 in)
Depth: 30 m (100 ft)
Site: Sarawak

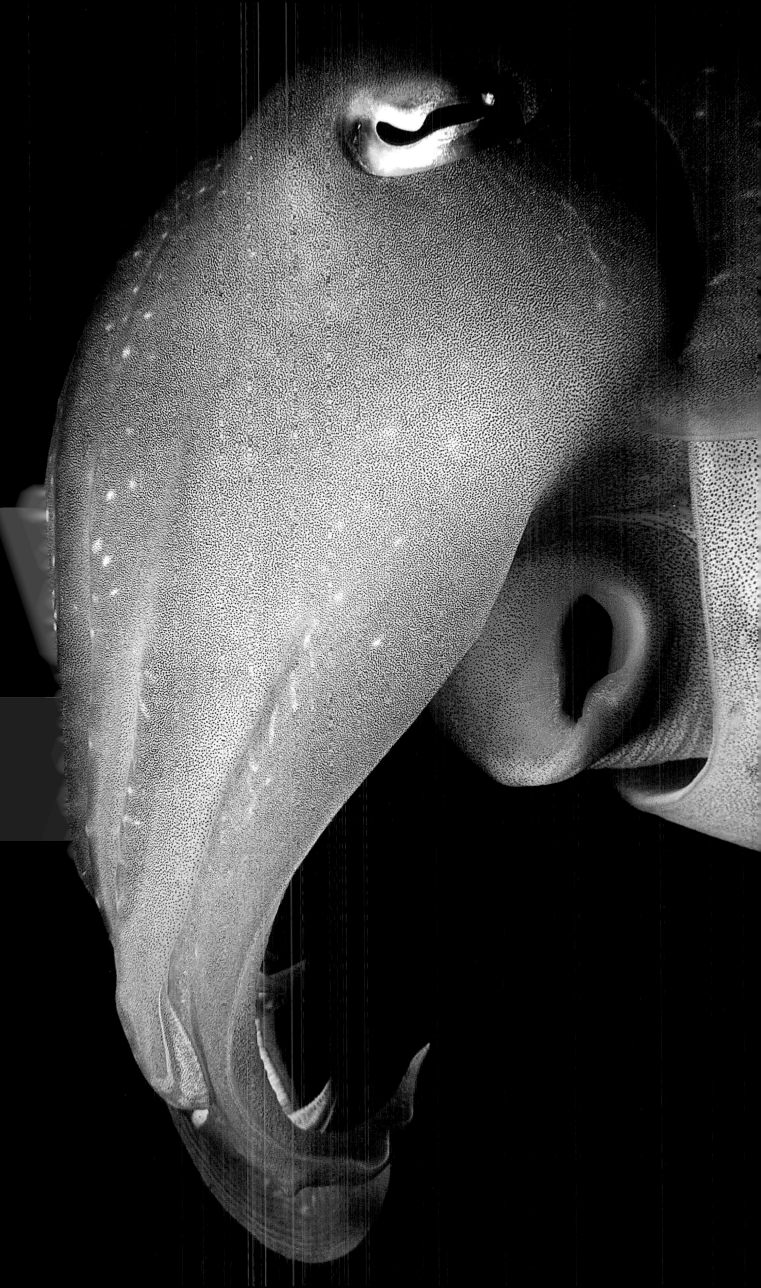

# THE PHILIPPINES

In these seas

*In the age of the dinosaurs,*

the variety of life

*it was here that evolution*

exceeds the imagination.

Palawan. A luminous sea. A dead leaf lies on the sand, in 6 metres (20 feet) of water. On edge, it moves in the current as though stirred by the wind in a clearing. It is red, oval and ribbed, its face thin, like an arrow. Some details catch the attention: an oblique slit suggesting a mouth, two black points calling to mind a pair of eyes.

A shrimp passes in front of it and is engulfed by what really was a mouth. The fishing technique of the leaf fish is effective, its camouflage skills consummate; a dorsal fin like a halo and movements that imitate a land plant gently swaying in the breeze. A trap for its prey and protection from its enemies. But why has a fish chosen to fake a fallen leaf?

There are plenty of dead leaves in the sea at Palawan: the nearby forest is lush and the breeze and the streams carry the leaves to the sea. The leaf fish (*Taenianotus triacanthus*) imitates what it finds. It is a cousin of the scorpion fish, and venomous, like all scorpaenids. It can be found in various colours — reds, yellows, black, white… These fish moult. From green they turn brown: just like the leaves.

Another Indo-Pacific species, from a neighbouring family (the Tetraorgidae), adopts the same type of disguise: it is called the cockatoo wasp fish (*Ablabys taenianotus*) because its forward-projecting dorsal fin is something like the eloquent erectile crest of an Australian parrot. A simple convergence of shapes… The photographer appreciates the subtlety of this deception. She prepares her flash, focuses, and takes a portrait of this frank dissembler.

### BIRTHPLACE OF THE CORALS

The Philippines… Sophie loves the reefs of this archipelago. First, simply because they are magnificent. Then, because they host the most prodigious variety of life imaginable. These waters are the fertile matrix and birthplace of today's corals. They conceal the evolutionary roots of our madrepores. It was in the Mesozoic era that modern corals appeared, after massive extinctions in the Permian (especially of the tabular and ridged corals), at the time when the Tethys sea split the primary supercontinent (Pangaea) into Laurasia in the north and Gondwanaland in the south.

The hexacorals with their polyps of tentacles in sixes, or multiples of six. Furthermore, in these waters can be seen most ancestral lines, classes or aquatic families — from sponges to mammals — in an incredible variety of shapes and sizes. Sophie disembarked at Puerto Princesa, the capital of Palawan, an island 450 kilometres by 25 (280 by 15 miles), to the west of the Philippines archipelago and separating southern China from the Sulu sea. The young woman has reached the villages in the west that guard Coral Bay and the Balabac Islands. Later she will head north towards Calamian, between Palawan and Mindoro.

**PRECEDING DOUBLE-PAGE SPREAD**
This little green turtle (*Chelonia mydas*) hatched during the night. It dragged itself across the sand and here it is in the sea, rowing towards its destiny…

Length: 7 cm (2.8 in)
Depth: on the surface
Site: Palawan

**ABOVE, LEFT**
The periclimenes shrimp never leaves its 'home, sweet home': the arms of an anemone!

Length: 2 cm (0.8 in)
Depth: 20 m (65 ft)
Site: Palawan

**ABOVE, RIGHT**
Organ pipe coral (*Tubipora musica*) spreads its eight-tentacled polyps.

Length: 1.5 cm (0.6 in)
Depth: 15 m (50 ft)
Site: Mindoro

**OPPOSITE PAGE**
These pedunculate clavelinid sea squirts (*Clavelina detorta*) demonstrate the ambiguity of the protochordates: simple in appearance, they heralded the arrival of the vertebrates — us!

Length of an individual: 0.5 cm (0.2 in)
Depth: 5 m (16 ft)
Site: Palawan

*The diver observes a leaf fish and ponders on the art of animal camouflage... Masks and pretences! Magnificent deceit!*

PRECEDING
DOUBLE-PAGE SPREAD,
LEFT
The closed mouth
of an elephant's ear
(Amplexidiscus fenestrafer).
This flattened
anemone unfolds its
disc to capture its
prey.

Length 8 cm (3 in)
Depth: 15 m (50 ft)
Site: Palawan

PRECEDING
DOUBLE-PAGE SPREAD,
RIGHT
The longnose hawk
fish (Oxycirrhites typus)
lies in wait among
clumps of hydras.
Its odd, squared
livery suggests a
tartan!

Length: 10 cm (4 in)
Depth: 40 m (130 ft)
Site: Palawan

ABOVE
The leaf fish
(Taenianotus triacanthus)
mimics a dead leaf
at the bottom of the
water.

Length: 10 cm (4 in)
Depth: 8 m (26 ft)
Site: Mindoro

OPPOSITE PAGE
The flatworms, or
planarians (genus
Pseudoceros), bring a
marvellous palette
of colours to the
reef. They are still
not fully known.
Above is a close-up
of Pseudoceros bedfordi,
and below a cousin
of Pseudoceros splendidus.

Length: 3 and 4 cm
(1.2 and 1.6 in)
Depth: 25 and 30 m
(80 and 100 ft)
Site: Mindanao

The diver submerges over a tongue of sand, then reaches the deepest part of the reef. She observes the leaf fish and ponders on the art of animal camouflage. *Masks and pretences*, to echo the title of a book on mimicry by Claude Nuridsany and Marie Pérennou, the authors of the film *Microcosmos*.

In the ocean, the forms of camouflage are legion. The simplest is also the most effective: it is used by creatures of the open sea. Two colours, the dorsum dark (black or blue-black) and the ventral surface white. Seen from above, the individual merges with the dark of the depths. Seen from below, it becomes as light as the surface illuminated by the sun. Many predators use this form of camouflage: sharks, tuna, jacks,

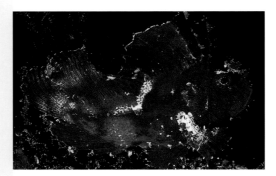

barracudas... Among the diving birds, cormorants and, in other parts, the penguins. Among the mammals, dolphins...

Other species exploit transparency, disappearing when the background can be seen through their bodies. Jellyfish, the ctenophores and the salps use this method. Shrimps just love to make themselves look like glass jewellery. Among the prettiest of these are the periclimenes shrimps: Sophie loves to portray their delicacy and photographs them like people. In the Philippines, the squat anemone shrimp (*Thor amboinensis*) relies on its bands of pink and grey-green to conceal itself in a similarly coloured anemone. And among the ovoid fingers of the bubble coral, the bubble coral shrimp (*Vir philippinensis*) presents its 2 centimetres (0.8 inch)

of white or beige light and its thin violet-mauve, stilt-like legs.

Another method of camouflage patented by Mother Nature is known as 'disruptive coloration'. Behind this abstruse expression there lies a basic truth: to avoid being seen, display yourself, show off and look good, on condition that you are covered in dots and patches, streaks, stripes or flames that interfere with your silhouette, break up your outline and make you unrecognisable. Become a complete optical illusion!

Many wrasse, butterfly fish, angel fish, damsel fish and surgeon fish look as though they have been cut in half – one half black, the other yellow: the front pink and the back green, and so on. Or they are broken up into checks, rectangles and zigzags... A predator no longer sees the whole animal, or not quickly enough. The livery of the cardinal fish (*Appogon nigrofasciatus*), with its lengthwise stripes in black and yellow, makes it difficult to see its 'fish' shape. The striped sweetlip, the blue-lined lutjan and the black-banded damsel fish play exactly the same trick. Sophie herself hesitates momentarily when she comes face to face with a clown trigger fish (*Balistoides conspicillum*), improbable in its make-up of yellow lips, black back, green-striped tail and its anthracite body with big white oval patches...

THE ART OF
DECEPTION

Under an arch of coral and sponges colonised by hydras and sea squirts, the photographer continues her search for beautiful masks. The primary function of camouflage is to disguise the most vital part of the body: the head and the eyes. The organs of vision are sunk into a dark patch: the animal displays false eyes (eye-spots) on its tail. It agrees to sacrifice the fringe of this appendage to save the essential. The enemy hesitates, then attacks, but strikes at the wrong part. Honour may have

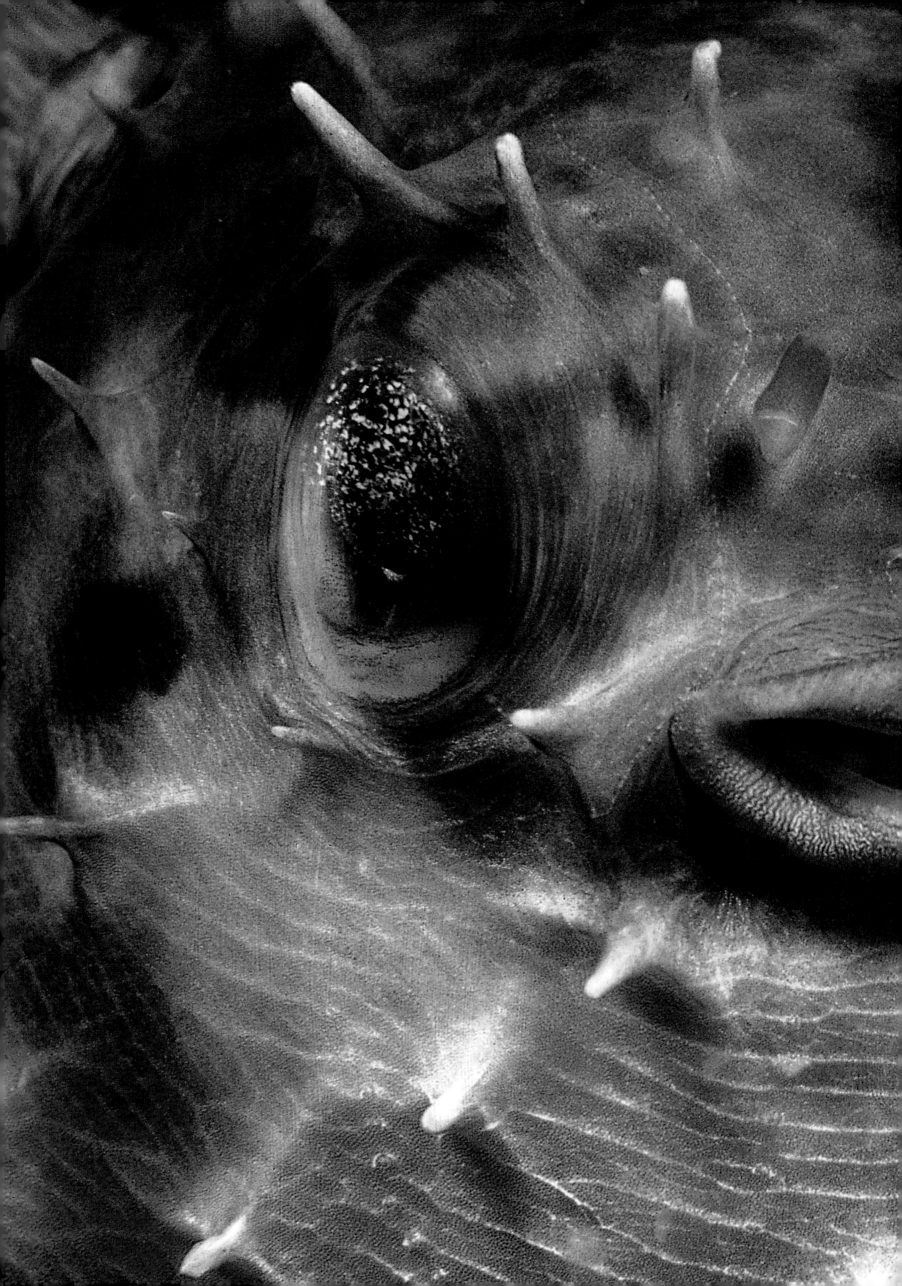

been lost, but not life. Other creatures adopt a 'commando' appearance. They put leopard spots on their uniforms, charcoal on their faces, and leaves on their helmets. Ghost and decorator crabs act in the same way. With the aid of its pincers, *Camposcia retusa* sticks grains of sand and the debris of algae and shells on its carapace and turns into what seems to be a piece of seabed on the march. The height of the art is to give oneself the same form and the same colour as the scenery. Better still to practise mimicry and imitate another species. Soles and crocodile fish, scorpion fish and stone fish are expert forgers. Their scales reproduce every bump, nuance and texture of their surroundings.

Before the diver's eyes, the crinoid squat lobster (*Allogalathea elegans*)

the mouth is large and the lower lip sloping. . If you are unlucky enough to be stung by this creature and are far from help, the most urgent thing is to cauterise the wound: better a nasty burn than death itself... Sophie swims in Palawan's spell. Attentive, fascinated, incredulous... A reminder of the meanderings of natural history, the inventions of evolution that produced these works by Bosch, Monet, Picasso, Pollock or Bacon, with their carapaces, spines, shells, scales, tentacles, pincers or teeth. On all sides there is nothing but illusions, sham, seduction and trickery. The diver descends to a seabed of hard coral acropores and soft coral *Dendronephtya*, rubbing shoulders with coneshells, murex and three species of sea-slugs in surreal colours: the

*On all sides, nothing but illusions, sham, seduction and trickery.*

displays the same shape and colour, and moves in the same way as the feathery arms of the sea-lily that is its home. The striped knife fish (*Aeoliscus strigatus*), transparent, and as thin as a blade, resembles the white eunicella gorgonias among whose branches a group now browses, head-down.

### BEAUTY TACTICS

The best camouflaged is the reef stone fish (*Synanceia verrucosa*): a fearsome scorpion fish! You do not see it. You put your foot on it, and the spines inject venom that is more effective than that of the cobra. Sophie has a good eye: she spots one of these innocent assassins near a mushroom coral. Its skin is grey-brown with bumps and fringes, with frayed bits of algae and sponges;

doris (*Chromodoris kuniei*), ochre and beige with a mauve border and violet spots; the Anne doris (*Chromodoris annae*), with three concentric grey-blue, black and orange bands; and the nembroth (*Nembrotha kubaryana*), in sinople (green) and sable (black), as a heraldist would put it.

Most of these shapes and colours are by no means the result of chance. They intrigue and charm us, but they are not intended for us, we merely observe them. We do not understand them. We are stupefied spectators, filled with wonder. Like Quasimodo confronted by Esmeralda, we are only able to repeat 'Beautiful! Beautiful!...'. Science may well try to explain the relationship between causes and effects, but in the end it fails. It falls to the poet to draw conclusions!

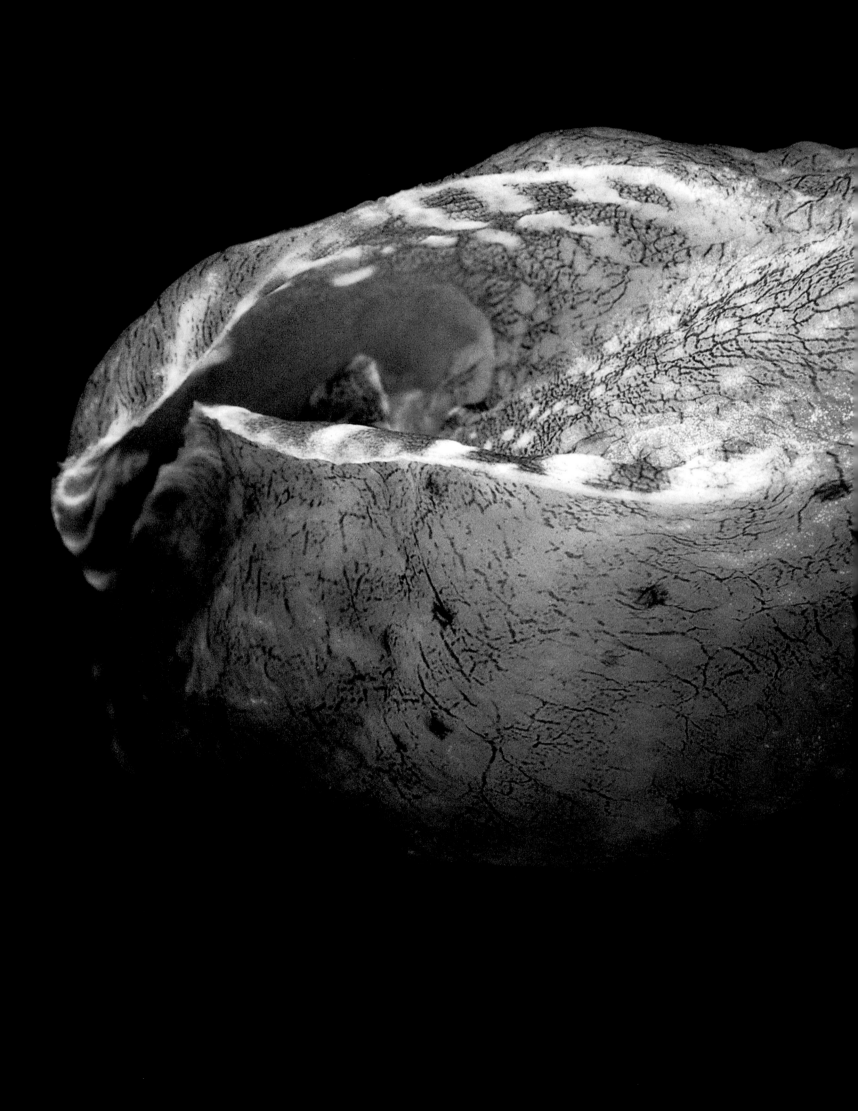

# AUSTRALIA

## The Great Barrier Reef

This is the largest reef system on earth.

## along the coast of

One night in the springtime:

## Queensland.

the start of an incredible event

It is snowing. Snowing in the sea…
But what strange snow! Snow in
reverse – the flakes swirl about as
they rise to the surface. Where is the
Ice Queen of this antipodal realm?
Where are the crib, the pine tree,
Santa Claus? This aquatic winter is
in fact due to the flight of millions
of reproductive cells… The coral
night of love is starting. Eggs and
spermatozoids soar, pirouette and
merge.

This mystery takes place by moonlight
in the waters of the Great Barrier
Reef, Australia. Sophie watches the
spectacle revealed in her spotlight.
She is diving at Cod Hole, actually
a 'grouper hole' – grey groupers
with black patches. The reef throbs
with a sexual rhythm. Along more
than 2,000 kilometres (1,250 miles)
of coastline, and over an area of
230,000 square kilometres (90,000
square miles; almost the size of the
United Kingdom), a sea of love.
The most imposing orgy on earth!
The Great Barrier Reef makes divers
dream. Blue. Sometimes, like
tonight, dark and starry like a dream
of Christmas reversed… the Eighth
Wonder of the World; a paradise of
turquoise and ultramarine splashed
with diamonds, rubies and sapphires;
an Eden where the trees are corals
and the bushes gorgonias; where
the angels are fish; the parrots peck
at the coral; the damsels and the
groupers take on rainbow hues;
where eagle rays and mantas glide;
where morays and sea-snakes
undulate; where the great white
and wandering tiger sharks slip by;

where the laughing dolphin clicks;
where sings the humpback whale
(*Megapter*) with its huge white
flippers…
The Great Barrier Reef is so precious
a natural monument that the United
Nations has pronounced it a World
Heritage Area. In keeping with the
vitality of the ocean itself, its
incredible inventive capacity, and
the splendour and strangeness of its
creatures the Australians have made
it into their finest national park.

## A NATURAL MONUMENT

In truth, the Australian Great
Barrier Reef is one of the planet's
most fabulous ecosystems. In terms
of its complexity and the diversity
of its species, it can be compared to
the Amazon jungles. It stretches from
the Gulf of Papua (in Papua New
Guinea) to Lady Elliott Island,
south of the tropic of Capricorn.
Its width varies from a minimum of
35 kilometres (22 miles) opposite
Melville to a maximum of
260 kilometres (160 miles) to the
right of Mackay. It is a complex of
some 2,500 barrier reefs, with
another 540 fringing reefs attached
to Queensland, plus another
200 islands farther out to sea.
The total surface area is impressive,
but the portion occupied by
madrepores is only about 20 per
cent of the whole. This leaves room
for the water.
This immensity, these castles,
temples, pyramids, villages and

PRECEDING
DOUBLE-PAGE SPREAD
The giant sea hare
aplysia *(Aplysia gigantea)*
grazes on coastal
plants and algae:
two pairs of wide
tentacles on a big
soft body…

Length: 30 cm (12 in)
Depth: 5 m (16 ft)
Site: Cod Hole

OPPOSITE PAGE
The fine striped
bubble shell *(Hydatina
physis)* is one of the
sea-slugs. It still has
a shell, but cannot
withdraw its body
into it.

Length: 4 cm (1.6 in)
Depth: 4 m (13 ft)
Site: Flinders Reef

ABOVE, LEFT
The striking tube
polyps of the tube
star *(Tubastrea sp.)*
deserve the name
of 'flower-corals'.

Length: 4 cm (1.6 in)
Depth: 35 m (115 ft)
Site: Pompey Reef

ABOVE, RIGHT
The nudibranch
*(Chromodoris kuniei)*
graces many reefs
in the Pacific
Ocean.

Length: 5 cm (2 in)
Depth: 25 m (80 ft)
Site: Osprey Reef

oriental palaces delight the five senses. The sight, naturally, since the corals and their guests present a profusion of shapes and colours, whose superabundance and staggering variety exclude neither harmony nor subtlety... A pleasure for the sense of smell, as soon as the nose leaves the water. For the sense of touch: the softness of the sand on the skin, the tickle of a sponge. As for the taste, it goes without saying, the salt of the Coral Sea has all the savour of existence. And finally, for the ear, the snap of the pistol shrimp's pincers; the agitated grunting of a disturbed hog fish; the explosive gaping of the potato grouper (*Epinephelus tukula*); the strange noise like crumbling biscuit produced by parrot fish as they graze on the coral; the clicking of dolphins, and the long, sometimes lowing, sometimes warbling notes of humpback whales.

### THE GREAT NIGHT OF LOVE

Sophie had already visited the Great Barrier Reef many times in broad daylight. She has been waiting for this night dive. Of course her best photos are not to be taken during the dark hours. She wants to survey the scene. She finds it hard to believe in this happening. The mass reproduction of corals is still a scientific mystery. The phenomenon begins — with sidereal accuracy — on the second night following the first full moon of the southern spring. In October or in November. As though at a signal, seeming to follow the orders of a conductor, all the corals of the internal reef begin to lay eggs. Together. Obsessively. Frenetically. A month later it is the turn of the madrepores on the external reefs. Why this urge, this synchronisation? The phenomenon has only been known about for a few years. No chemist, biologist or ecologist has been able to answer the bulk of the questions it poses. The tides play some part, because the moon is

involved: but why these tides in particular and not others? One imagines it must be a combination of different trigger factors — the tide, the relative lengths of days and nights (circadian rhythm), the water temperature, etc. Do the madrepores also emit subtle chemical messengers — pheromones — that excite their sexual partners? So far no such substances have been identified.

### A POLYP GIVES BIRTH

Sophie swims down to a montipora finger coral. A mineral bush bristling with delicate antler-like branches, creamy-white with hints of mauve at the forks. Hundreds of polyps with pink arms are on display, a flowering shrub. Sophie observes, examines and with her fingertips strokes the

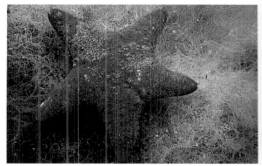

crown of an individual coral, one in the colony. At the centre of the whorl of tentacles, the mouth opening, the walls of the tube are swollen, turgescent, and seem a brighter crimson than usual. The opening widens. Our diver watches as a whitish mass emerges. It is as though the polyp were giving birth to this packet of eggs in silent suffering. The mass poises, then frees itself and rises towards the surface like some miniature hot-air balloon. A second packet of cells begins its ascent. Some other globules follow. At this moment, all the polyps are laying eggs. A council of love. A storm of life. Snowfall rising into the sky: a gift from the sea! As the eggs are expelled into the water, the polyps emit jets of

*The diver observes, examines and photographs the whorl of a coral. These corals and their improbable world are a mystery...*

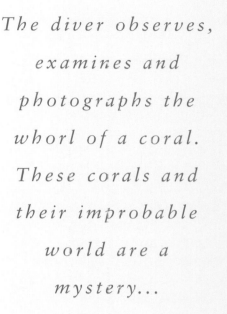

**PRECEDING DOUBLE-PAGE SPREAD, LEFT**
This colony of ascidians of the Styelidae family *(Distomus sp.)* carpets a block of dead coral. It is difficult to see our ancestral cousins in these sacs!

Length of one sac: 1 cm (0.4 in)
Depth: 30 m (100 ft)
Site: Cod Hole

**PRECEDING DOUBLE-PAGE SPREAD, RIGHT**
The flying lion fish *(Pterois volitans)* combines splendour with venom: such is the beauty of the devil!

Length: 30 cm (12 in)
Depth: 45 m (150 ft)
Site: Ribbon Reef

**OPPOSITE PAGE, ABOVE**
A giant spirobranch worm *(Spirobranchus corniculatus = S. giganteus)* spreads its double plume of tentacles on a diploastrea star coral.

Length: 3 cm (1.2 in)
Depth: 35 m (115 ft)
Site: Cod Hole

**OPPOSITE PAGE, BELOW**
The ventral face of a marble sea-star *(Fromia sp.)*... members of this echinoderm order eat by everting their stomach through the mouth.

Length: 5 cm (2 in)
Depth 15 m (50 ft)
Site: Osprey Reef

**ABOVE**
One of the innumerable stars in the sky of the sea on the Great Barrier Reef: this one is a juvenile of an unknown species.

Length: 5 cm (2 in)
Depth: 20 m (65 ft)
Site: Swain Reef

iridescent milky sperm in wreaths, twisting in the light. At the whim of the currents, millions of fertilisations take place below the surface. Species other than the montiporas take part in the orgy.

## FERTILISATIONS BY THE BILLION

Biologists have identified acropores such as *Acropora tenuis*, *A. formosa*, *A. humilis*, *A. valida*, etc. Porites, coral-flowers, tube corals, diploastreas, turbinarians, oxypores, bubble coral, platygyres, mushroom coral (*Pavona*, *Fungia*), club coral, dendrophylls, etc. Eggs and spermatozoa meet and pass, mix and merge haphazardly.
The photographer swims in water that has become the soup of a

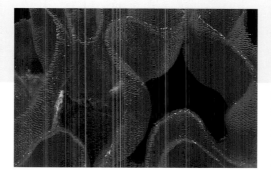

fantastic gene lottery. The 'night of love' actually lasts for little more than an hour.
The following day and for four nights the madrepores will endure a fresh sexual crisis. The polyps that have not laid eggs tonight will be given another chance in a month's time, at the next full moon; then it will be over for the year. Nine-tenths of madrepores perpetuate their species in this way, in a few hours of springtime madness. The mystery of nature's major meetings.
The vast majority of eggs and spermatozoids are devoured by the millions of shrimps, squid and little fish that hasten to the banquet table (how they know is another mystery). About an hour and a half after fertilisation the surviving eggs turn orange, violet or blue. They divide

into two, four, eight, sixteen, etc. In 48 hours, a planule, a flattened larva with cilia hatches out, characteristic of the coelenterate line. This animalcule, which measures 2 to 3 millimetres (about 0.1 inch), is a part of the plankton. If it manages to escape from predators (the chance of survival is small), it allows itself to drift. If it finds a tiny area of free space on the bottom, it metamorphoses into coral. It develops a calcium carbonate skeleton: a base plate and a tiny cylindrical wall, reinforced by six panels called septa pointing towards the centre. Round this structure the creature develops a calyx and a crown of tentacles.
The original polyp reproduces by growing. By cloning. It copies itself. It gives rise to male or female polyp children, which in turn produce grandchildren, and so on. Each individual, identical to its predecessor, builds a part of the monument. On average the polypary grows by 5 millimetres (0.2 inch) a year; 50 centimetres (20 inches) in a century. It touches other madrepores, sponges and gorgonias... It struggles for space and light. When it dies, its frame becomes the support for other beings. In this way, little by little, titanic structures arise.
Corals embody the patience of water. They are architects of genius. The pharaohs of the sea... Sophie continues through the water lit by the mauve and pink of dawn. A dolphin skims past, blowing. 'There all is order and beauty, richness, calm and pleasure...' Never have the accents of Baudelaire's *The Invitation to the Journey* found better expression in reality. The Great Barrier Reef is a vast poem of sensual delight.
It remains to be seen whether we humans will be able to preserve for future generations this inheritance from an evolution that the passage of millions and millions of years has made so beautiful, so necessary. The coral larva does not have the answer!

*The Great Barrier Reef: a fabulous ecosystem, a sensual poem...*

PRECEDING DOUBLE-PAGE SPREAD
These biscuit starfish *(Tosia queenslandensis)* have climbed onto a bed of feathery hydras, and seem to be enjoying a meal of polyps...

Length: 5 cm (2 in)
Depth: 15 m (50 ft)
Site: Heron Island

OPPOSITE PAGE
An ornate ghost pipe fish *(Solenostomus paradoxus)*, cousin to the seahorses, mimics a gorgonia, its red body bristling with false 'polyps'.

Length: 5 cm (2 in)
Depth: 35 m (115 ft)
Site: Osprey Reef

ABOVE
A detail of the spiral of eggs laid by the little Spanish dancer sea-slug *(Hexabranchus sanguineus)*.

Length: 3 cm (1.2 in)
Depth: 15 m (50 ft)
Site: Pompey Reef

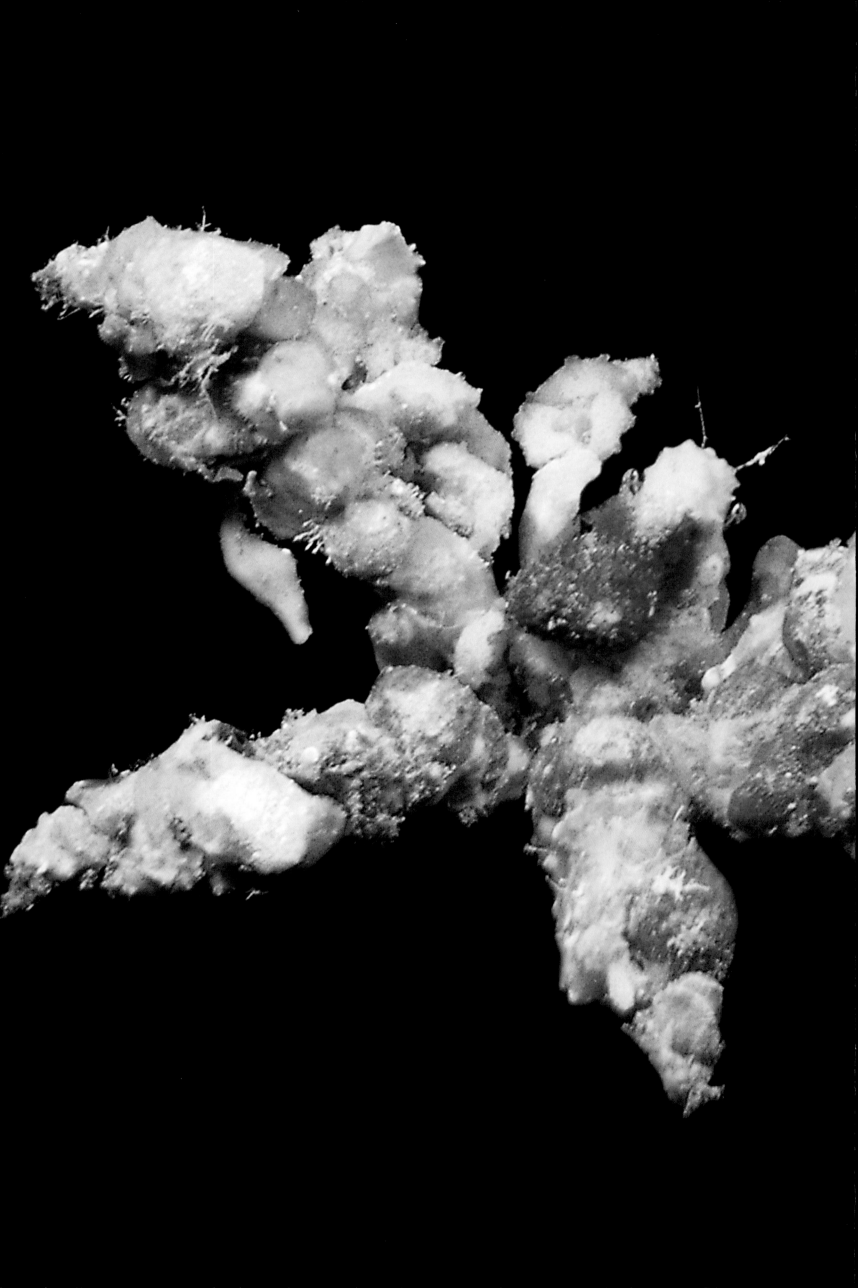

# TONGA

## The grace of the

*Tonga, Fiji: distant places that elicit*

## southern seas, amid atoll

*the bittersweet feelings experienced by*

## of ultramarine and pearl

*Gauguin and Brel in the Marquesas.*

# TONGA

Tonga. Lifuka island, in the Ha'apai group, between Vava'u to the north and Tongatapu to the south... Coconut palms murmur, the beach sparkles. Beyond the calm of the lagoon and the foaming reef, the ultramarine melts into flights of white birds. The scent of gardenias wafts in the breeze. Tonga... A sense of eternity ('the sea mixed with the sun' — Rimbaud); but a symbol of the transient. These coral islands will return to the sea that made them one day, but humans, treating them with contempt, ruin them much more rapidly.

Sophie immerses herself in the strong current of a channel where dolphins delight. She lets herself drift towards the relative calm of a ledge of madrepores exuberant with life. On a montipora like Mont Blanc, the diver sees one of those large brownish starfish called 'crown of thorns', or, less tactfully, 'mother-in-law's cushion'. An acanthaster (*Acanthaster plancii*) with 13 arms (the number varies with the individual) and a diameter of 60 centimetres (24 inches)... The echinoderm has found its prey: the montipora. It embraces its victim like a Mafia godfather: to assassinate it. It pushes its stomach out through its mouth and clamps it on the polypary. The gastric juices do their work, and the star digests the tender polyps.

## THE LITTLE HYMENOCERE MONSTER

The 'cruelty' of nature! Let us shed a sentimental tear at this Tongan drama! Let us complain about the coral being eaten alive  We are that much less inclined to defend the spiny starfish because it manages to proliferate. Gatherings of this predator have filed down, erased and wrecked some reefs, especially at Maurice island and in the Maldives — to the point where intervention to eradicate the 'scourge' has been considered. Is there no pity for the acanthaster? Let's see... Apart from ecology, affective criteria have little relevance. The following episode may temper the indignation. A little being trots along on its ten legs; at least on the six it uses for walking; the two others are armed with pincers. This is a crustacean with a carapace illuminated like a medieval manuscript. Mauve clouds on an alabaster ground... Sophie identifies a hymenocere shrimp: a painted shrimp (*Hymenocera picta*) or a blue harlequin shrimp (*H. elegans*): it is difficult to decide at a glance. The little beast, 8 centimetres (3 inches) long, climbs onto the starfish... and attacks! With blows of its pincers and jaws it slashes, cuts and bores into the skin. On reaching the flesh it begins to feed. It cuts the inside of the echinoderm to shreds, in just the same way as the latter digests the coral. It is a gory scene, but worse is to come .. Once it has devoured the starfish arm on which it stands, the shrimp moves on to the next, and the next... Since the shrimp is small and the starfish large, there is enough for weeks. But it has a good appetite. It never eats the central disc, or

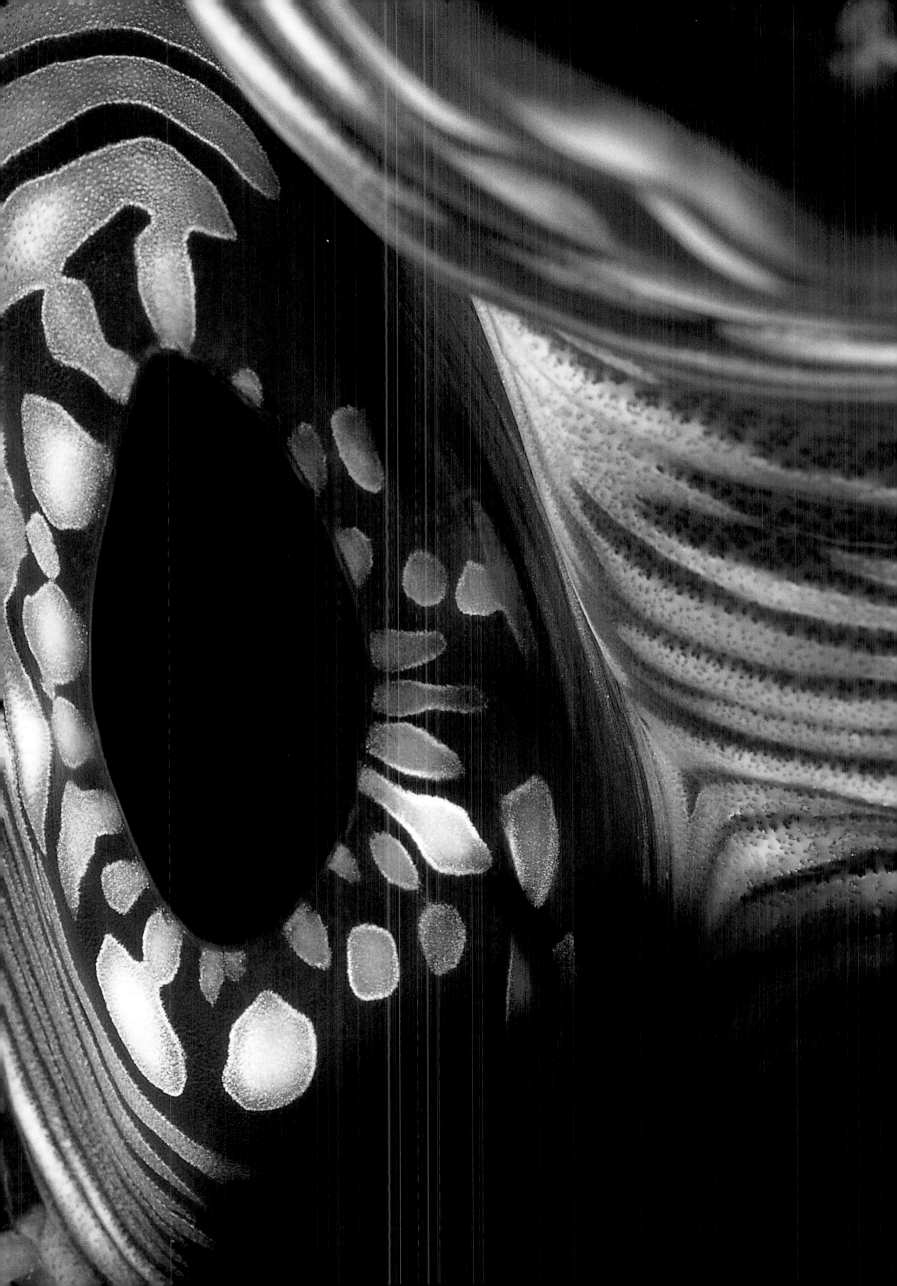

the last arm. The starfish only needs one arm to be able to regenerate all the others. Now all is clear: the hymenocere shrimp waits for the starfish to re-create itself to be able to eat it again!

The Tonga islands continue the Kermadec archipelago to the north, in line with New Zealand. In latitude they lie on either side of the tropic of Capricorn. In longitude, they straddle 175°W, close to the dateline. Sophie ponders on their geological origins. The Tonga islands are a part of the Ring of Fire of the Great Ocean: they are the children of volcanoes and the subduction of the Pacific plate under that of Australia. Some miles to the east there begins a vertiginous blackness plunging 10,882 metres (over 35,000 feet) below the surface (in the Horizon

Marseilles during the Second World War were copied from Polynesian goggles made of tortoiseshell, called *titia* by the locals. The photographer returns to the turbulent water in the channel. The current carries her along. The dolphins and sharks are more at ease... The play of tides between ocean and lagoon is the atolls' breath. The diver drifts, reaching a coral temple. A fish leads her to the altar: a humphead wrasse (*Cheilinus undulatus*), a debonair giant, sometimes 2 metres (more than 6.5 feet) long and weighing 200 kilos (440 lbs), with a blue-green body and fine worm-like ripples on its sides. This is a colossus of the Labridae family. Some of its cleaner fish measure at least 5 centimetres (2 inches). It has enormous lips and tiny eyes like boot buttons. And in

*Dungeons, ramparts, towers...*
*After the gardens of Babylon, the palaces of Egypt...*

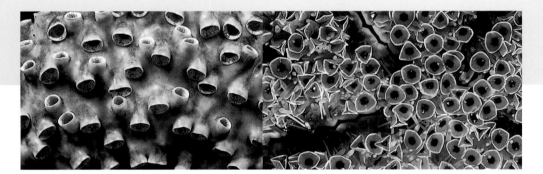

Fault of the Tonga-Kermadec system). This is the second largest 'depth peak' on the earth's crust — after the Challenger Deep in the Marianas Islands.

As humans who dive just a few dozen metres, encumbered with equipment, it is difficult not to think about these abysses ruled by absolute darkness and tremendous pressures; scarcely known to oceanographers, home to the most fantastic creatures in the liquid universe.

### THE EBB AND FLOW OF LIFE

Sophie takes in a little air from her mask. She remembers a meeting with the chief diver of Cousteau's crew, Albert Falco. He told her that the first diving goggles he tried at

front, the emperor's hat of a hump that gives it its name...

Sophie moves behind the huge animal: it does not fear people. But it pays for its placid nature. Thought of as a trophy by some sad hunters with spear guns, it is becoming rarer. Biodiversity in danger!

From amid the branches of an acropore, our diver contemplates the pillars, ramparts, dungeons, towers, crenellations and blocks of reef. Ancient palaces give way to the gardens of Babylon. One can recognise imitations of Borobudur and Reims cathedral. The White House and Tiananmen Square... Elsewhere one imagines Polynesian thatched houses, suburban houses or low rent housing. The architecture is varied. All styles are represented. But if nature prizes excess, it never

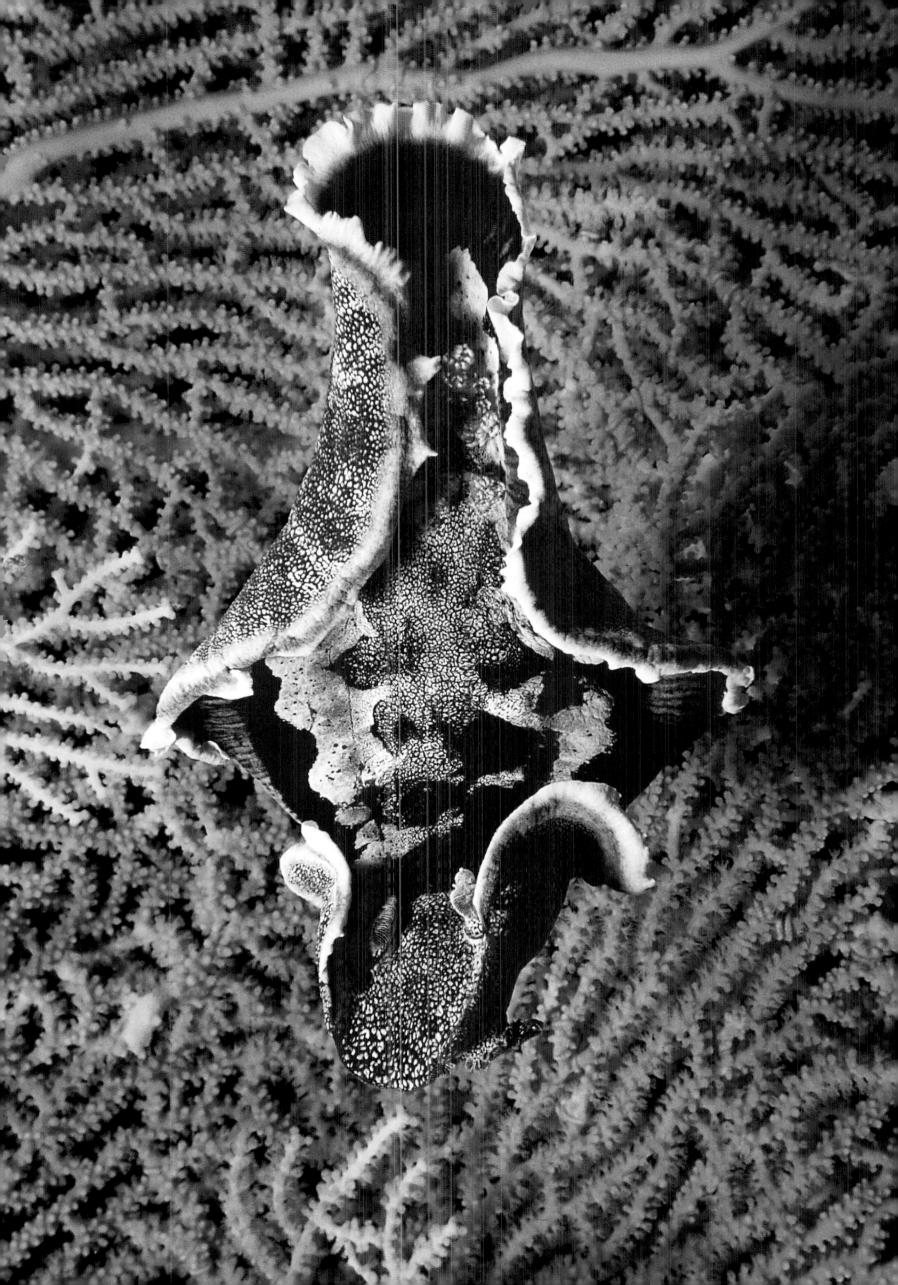

commits errors of taste. Once again
it falls to Sophie to admire the variety
of building organisms. Branched or
brain-shaped coral; in beehives,
mushrooms, plaques, globes, salads,
tables, draperies and what else?
In tablemats, plates and slippers!
Life's imagination overflows. The
corals are not alone in depositing
their strongholds and Versailles of the
sea, millimetre by millimetre. The
algae make their contribution too.
For example, the halimeda (*Halimeda
opuntia*), with its fronds in strings of
bottle green triangles. Or the red

encrusting algae that create harder,
more compact carbonate structures
than the madrepores. These poroliths
and lithothamnes affix their pink, red
or pomegranate plaques, leafy, in
waves or horns to the most exposed
edges of the reef, where the surf
breaks with the greatest fury.
Other algae play a decisive role.
The building corals feed partly on
plankton caught in the current by
their venomous tentacles. But most
of their energy comes from glycerol
(a compound related to the sugars)
provided by minuscule brown
unicellular algae, zooxanthellae,
that colonise even the tissue of
their polyps. These latter (*Symbodium
microadriaticum*) also benefit from
communal life: in the coral they
find shelter and the nitrates and
phosphates indispensable to their
growth and reproduction.
The symbiosis of the zooxanthellae and
the madrepores explains why these last,

who live exclusively in warm waters (at least 18°C/64°F), also require light: that is, clear water. When the turbidity is too great (at the mouths of rivers or in areas polluted by humans), the lack of photons prevents the algae from photosynthesising. The corals stop growing, secreting calcium carbonate, and reproducing — and die in a few weeks.

## FRAGILE SPLENDOURS

Imposing yet vulnerable marvels! Everywhere on the planet the reefs are threatened. Not only do we break them up to build houses, to provide stone for roads or when excavating ports, we also choke them by moving the earth and cutting down forests on land bordering the sea. We break coral up by fishing with dynamite or using trawl nets. We poison them with cyanide or fuel oil. Perhaps the greenhouse effect and the changing

climate for which we are responsible will be their deathblow. They suffer from the raised sea temperature and excessive ultraviolet rays that result from the thinning of the ozone layer. They expel their zooxanthellae and whiten. A strange and disturbing suicide. A symbol of the fate in store for our species?

The photographer caresses an acropore with grey-pink horns. Like the estuaries, the littoral meadows, the marshes, the mangrove swamps, the zones where currents meet, or where deep waters rise, the coral reefs are luxuriant oases in the marine desert. They are miracles in reprieve. On a yellow sponge, an ophiuroid — a brittle-star — twists its five fragile arms. (The Polynesians call this species *mcamatai*). The animal doesn't care. It lives. Simply. Innocently. Without asking questions. 'Without why', like the rose of Angelus Silesius. But for how long?

PRECEDING
DOUBLE-PAGE SPREAD
The long-legged periclimenes shrimp *(Periclimenes brevicarpalis)* or 'peacock's tail' is a table companion to such sea anemones as the giant *Stichodactyla<*.

Length: 3 cm (1.2 in)
Depth: 15 m (50 ft)
Site: Vanua Levu Island, Fiji

OPPOSITE PAGE, ABOVE
This mass of giant gorgonias *(Subergorgia mollis)* makes a fertile sunken copse where dozen of species coexist and flourish.

Length of a fan: 1.5 m (5 ft)
Depth: 25 m (80 ft)
Site: Lifuka Island

ABOVE
A black variety of Bennett's feather star *(Oxycomanthus bennetti)* spreads its plumes over an orange knotted gorgonia *(Melithaea sp.)*.

Length: 25 cm (10 in)
Depth: 40 m (130 ft)
Site: Vava'u Island

OPPOSITE PAGE,
BELOW
On the sand in shallow water, an invasion of blue starfish *(Linckia laevigata)*.

Diameter of a starfish: 20 cm (8 in)
Depth: 2 m (7 ft)
Site: Tongatapu

# CALIFORNIA

With a brief detour to the

*Gladly astray in the kelp forests*

Galapagos, the world diving

*with the sea otter, the cardinal fish*

tour ends in California.

*and the ghoulish wolf fish.*

# CALIFORNIA

The sea otter — wicked eyes, white face, black nose, moustache, fur cloak... A Californian star! Don't disturb her, she sleeps on the ocean, rocked by the swell, swathed by an alga as though in a green satin wrap. California... Many dreamed of it, some reached it. Under water it is ignored. The ocean depths of this land that was a dream of bonanza for millions of men hides much more than nuggets. They present a surprising and unusual theatre: entrancing scenes that encourage meditation and poetry... Gleaming displays spilling over with precious stones, splashed with pure colour. Sophie always preferred dream-like spectacles charged with mystery, glinting, shimmering, a play of mirrors in some fantastic fairy story.

California beneath the waves seems to owe nothing to the flashy have-you-seen-me world of Hollywood. In the Pacific the water is not swimming pool blue but a silvery blue-green and full of particles. Here wealth is not counted in dollars, except for sand dollars, the little sea urchins also known as 'clypeaster' or 'encope' (*Encope californica*). Power is measured not by the horsepower of noisy cars on a motorway but by the movements of a sea lion on the shore or the energy of a grey whale ready to leave on a 8,000 kilometre (5,000 mile) journey.

When oddity and cells combine to produce beauty, it's a miracle... Such is the elegance of life. Sophie

dons a thick diving suit for the water is cold. (A few days before she was in the Galapagos.) She makes herself ready to dive at Lobos Point, near Monterey, about 200 kilometres (125 miles) south of San Francisco. The Pacific rollers shatter on the coast. At a cable's length from the shore, sea otters are resting on the surface, draped in algae, floating on the swell.

These animals (*Enhydra lutris*) have nearly disappeared in California, as in Siberia and Alaska, because they have beautiful fur — the finest, according to trappers. Having crossed the Bering Strait and descended the west coast of North America, the Russians pursued the species as far as this, arriving in 1812.

By now the water nymphs of California are protected. They have reconquered their realm. They can be met with close to Los Angeles in some of the Channel Islands, where six species of pinniped have also made their return: the northern sea elephant (*Mirounga angustirostris*) a miraculous recovery, the Californian sea lion, the Guadaloupe fur seal, Steller's sea lion, the common seal and the bearded seal.

## TASTY SHELLFISH

The sea otters suffer from pollution by humans, most particularly with PCBs, heavy metals and fuel oil. Some fishermen still hate them and kill them out of revenge: they accuse them of devouring 'their' shellfish and crustaceans. Close to Monterey,

PRECEDING
DOUBLE-PAGE SPREAD
The Californian sea lion *(Zalophus californianus)* inspects the diver: mutual curiosity. Definitely a shared feeling for beauty under the sea.

Length: 2 m (7 ft)
Depth: 10 m (33 ft)
Site: Galapagos

OPPOSITE PAGE
The end of a starfish arm *(Astropecten sp.)* The small white tentacles are its walking legs, used for climbing about.

Length: 4 cm (1.6 in)
Depth: 20 m (65 ft)
Site: Channel Islands

ABOVE, LEFT
A meeting of violet sea urchins *(Arbacia sp.)* on the bed of a kelp forest — voracious consumers of algae!

Length: 7 cm (2.8 in)
Depth: 30 m (100 ft)
Site: Monterey

ABOVE, RIGHT
The bat star *(Patiria miniata)* advances on a coralline alga, looking for clams and mussels.

Length: 9 cm (3.5 in)
Depth: 35 m (115 ft)
Site: Channel Islands

the troop of otters has finished its siesta. The water nymphs dive in search of the invertebrates they love: abalone (*Haliotis rubescens*, *H. cracherodii*), clams (*Tresus nuttali*), crabs, lobsters, sea urchins... They know how to open the shells and break the carapaces of their prey with a tool: a stone picked up from the seabed and placed on the chest as an anvil. Sophie breaks through the surface barrier and drifts down into the splendour of the submarine forest. As in the Falkland Islands, this is a community of giant brown algae,

particularly 'pear-bearing' macrocysts (*Macrocystis pyrifera*). There is nothing as majestic as this underwood, on whose fronds the light plays as in a Corot painting. Californian macrocysts can be compared to the thuya trees, sequoias and colossal cedars that clothe the neighbouring mountains.

### THE CALL OF
### THE FOREST

Sophie caresses the bulbous bladders that serve the algae as floats. Some little pineapple fish swim among the fronds. Pink or yellow sea-snails climb the lustrous stems, thick, solid, somewhat sticky.
A shadow: a sea otter. The pretty carnivore swims down the length of a macrocyst, turns round the 'trunk', roots about in the holdfast and flushes out a crab: the promise of a feast. Like the otter Sophie

slaloms between the stipes. There is something unreal about this outing. The light falling from the surface is constantly changing, yellow-green or blue-green, like that in a forest on land, when the wind begins to blow.

The kelp ecosystem is fabulously rich. Thousands of species can be counted therein. The diver lists and photographs some examples. Take these bryozoans draping a dead gorgonian with lace; black and white scales chiselled by a tiny imp; it is hard to say whether naturalists have described this species. In these matters, hesitation is the rule and certainty the exception. Sophie would often take a picture of a species, then try in vain to name it, even in consultation with the world's greatest experts in oceanographic museums. Violet or mauve sea urchins scale the brown algae: these are kelp grazers. If their population were to explode

the plants would be over-grazed, and the collection of creatures that depends on them (including the abalones, the clams, the crabs and the lobsters harvested by the fishermen) would suffer. Now, who controls the demography of sea urchins? The sea otters do. And the fishermen accuse them of destroying a resource they actually preserve.

## THE SENILE ANEMONE AND THE GARIBALDI FISH

A chilly current rises from some unknown canyon. Sophie swims down towards the seabed. The rocks bristle with powerful sea anemones, whose livid tentacles remind one of an old man's beard. They are called 'senile anemones' (*Metridium senile*). Other actinia such as *Anthopleura xanthogrammica* are neighbours to hydras, sea-feathers,

PRECEDING DOUBLE-PAGE SPREAD
This Californian spider crab *(Maja* sp.) carries a sea anemone on its back that has vomited pink defensive threads.
A reciprocally beneficial association.

Length: 15 cm (6 in)
Depth: 25 m (80 ft)
Site: Channel Islands

OPPOSITE PAGE, ABOVE
The tiny equine outline of many different seahorses can be seen in California. This odd weedy sea dragon *(Phyllopteryx taeniolatus)* is from Australian waters.

Length: 20 cm (8 in)
Depth: 15 m (50 ft)
Site: Kangaroo Island

ABOVE
The Californian octopus *(Octopus californianus)* will not remain in open water: every octopus is obsessed with the idea of finding a safe hiding place.

Length: 40 cm (16 in)
Depth: 30 m (100 ft)
Site: Monterey

OPPOSITE PAGE, BELOW
The red-lipped bat fish *(Ogcocephalus* sp.) did not steal its nickname 'lips-on-fire'!

Length: 20 cm (8 in)
Depth: 20 m (65 ft)
Site: Channel Islands

*California beneath the waves has nothing in common with the flashy, have-you-seen-me world of Hollywood.*

the pink balls of sea squirts, and seashells: the tulip with its scarlet foot, the white murex, the spotted cowrie. A bright red-orange fish rises from behind an alga and darts towards the diver: a garibaldi (*Hypsypops rubicundus*). About 30 centimetres (12 inches) long this damsel fish named after Giuseppe Garibaldi, the leader of the 'redshirts' in Italy, is among the most territorial of fish. It defends its nest with unstoppable ferocity. Risking death, it even defies the sea lion. Sophie remembers an experiment tried by another diver: placing a mirror in front of its home. The irascible owner attacked its own image with such murderous intent that the test had to be called off. Schools of loligo squid and fork-tailed jack crevalles swim between

hides beneath a rock: no luck, an octopus lay in wait. Sea-slugs in crazy colours: yellow-green chromodoris with lettuce green spots, Dironidae in milk and silver, Polyceridae in pale yellow with blue bands, scarlet flabellinids with gold-flecked fingers... Not forgetting the surprising lemon sea-slug (*Cadlina luteomarginata*), which not only has the colour of the fruit it copies, but also secretes the same smell to repel its enemies.

Another perfect torpedo: a grey and white dolphin with a short, black muzzle, a white-flanked *Lagenorhynchus*. The killer whale could not tarry. Nor could the migrating grey whale. Sophie explores the opening of a cave in a wall of rock. A ghostly head shows itself at the window: small sunken eyes, a leprous

**PRECEDING DOUBLE-PAGE SPREAD**
One of Sophie's favourite photos: a sea lion (*Zalophus californianus*) charging in open water.

Length: 2 m (7 ft)
Depth: 30 m (100 ft)
Site: Galapagos

**ABOVE, LEFT**
This sea lion (*Zalophus californianus*) seems to glide at a shoal of anchovies.

Length: 2 m (7 ft)
Depth: 30 m (100 ft)
Site: Galapagos

**ABOVE, CENTRE**
A roving Galapagos shark (*Carcharhinus galapagensis*), more curious than aggressive.

Length: 2 m (7 ft)
Depth: 10 m (33 ft)
Site: Galapagos

**ABOVE, RIGHT**
This Galapagos penguin (*Spheniscus mendiculus*) takes advantage of the cold Humboldt current to live below the equator.

Length: 30 cm (12 in)
Depth: 5 m (16 ft)
Site: Galapagos

**OPPOSITE PAGE**
There is only one sea... the world ocean is an immense organism in which all creatures depend on each other and are of equal importance – from plankton to the sea-slug, to the whale, or to these Californian sea lions.

Length: 2 m (7 ft)
Depth: 3 m (10 ft)
Site: San Francisco

the stems. A small beige horned shark with black spots follows its angular muzzle. Sizeable shadows wheel about the diver: Californian sea lions (*Zalophus californianus*), the height of elegance. In the Galapagos, Sophie photographed this pinniped in movement: it was almost a self-portrait. A complicity. A communion.

### THE GHOULISH WOLF FISH

Some sea cucumbers with plumes of branched tentacles crawl through communities of sea-suns (*Heliaster*), yellow spotted stars and spiny sea-stars. Sophie examines the bulky *Patiria miniata* starfish, with its short arms and overgrown, raised central disc. A Californian spiny lobster

nose, deathly pale, a mournful mouth armed with big teeth at all angles... This is a wolf fish (*Anarrichas lupus*), two metres (7 feet) long, with a body that finishes in a rat's tail. Despite its appearance it is harmless and, despite its air of sadness, happy with its life. Sophie looks at this ocean ghost. One ending is like another in this tour of the world with flippers... She hopes that the time will come for humanity to follow the message of the German poet Novalis, *The Apprentices of Sais*, 'Let anyone who has the right feeling for nature profit from its study; rejoice in its infinite complexity and the abundance of its inexhaustible pleasures. There is no need for anyone to come and spoil his enjoyment with useless words.'

Gulf
of
Mexico

Tropic of Cancer

Antilles
Sea

ATLANTIC

Equator

OCEAN

Tropic of Capricorn

OCEAN

Mediterranean
Sea

Red Sea

Oman
Sea

Gu
o
Ben

INDIAN

OCEAN

| | | |
|---|---|---|
| 1 Falkland Islands | 10 Kerry | 19 Israel |
| 2 Honduras | 11 Costa del Sol | 20 Egypt |
| 3 Belize | 12 Costa Brava | 21 Sudan |
| 4 Florida | 13 Gulf of Lions | 22 Djibouti |
| 5 Bahamas | 14 Corsica | 23 Tanzania |
| 6 Dominican Republic | 15 Sardinia | 24 Aldabra |
| 7 French Antilles | 16 Tunisia | 25 Madagascar |
| 8 Grenadines | 17 Sicily | 26 Seychelles |
| 9 Connemara | 18 Aegean Sea | 27 Reunion |

*PACIFIC*

*OCEAN*

*South China Sea*

*Philippine Sea*

Galapagos Archipelago

*Coral Sea*

Fiji

Tonga

Tuamotu

*Tasmanian Sea*

28 Mauritius
29 Rodrigues
30 Laquedives
31 Maldives
32 Sri Lanka
33 Thailand
34 Sonde Islands
35 Sulawesi
36 Kalimantan

37 Sabah
38 Palawan
39 Mindanao
40 Great Barrier Reef
41 Kangaroo Island
42 Papua New Guinea
43 Vanuatu
44 New Caledonia
45 Fiji

46 Tonga
47 French Polynesia
48 California
49 Galapagos
50 Costa Rica

# DIVING NOTEBOOK

## By Ariel Fuchs

In some twenty years of diving, Sophie de Wilde has completed a real world tour. Here are the main places she has visited and photographed.

### 1. THE FALKLAND ISLANDS
Situated off the Argentinian coast, the Falkland Islands are not strictly speaking a classic diving location. Reserved for adventurers keen on subantarctic fauna and flora: there are no reception facilities, and any logistic support required must be taken along. The best season to explore them is naturally the southern summer, from December to February, but this does not mean you can dispense with waterproof clothing. Special feature: forests of kelp populated by strange, colourful invertebrates. Access from Britain is via RAF Brize Norton; elsewhere, from Santiago, Chile.

### 2. HONDURAS
The Bay Islands off the north coast of Honduras form a group where the world of coral is mixed with volcanic rock pierced by caves and other rifts. Exploration is possible the whole year round, but the dry season from December to March is best for an appreciation of the giant sponges, fan corals, vivid coral and the stingrays of Roatan. An exclusive, well-organised location. Access via Spain or the US. Not to be missed: an appointment with the bottlenose dolphins of Anthony's Key – an unforgettable meeting.

### 3. BELIZE
South of Mexico, off the coast of Belize, is the second largest coral barrier in the world. The reefs are of the utmost beauty, especially around Lighthouse Reef, Glover's Reef, Southwater Cays and Turneffe, with richly decorated drop-offs inhabited by an abundant, multicoloured fauna. One can dive there all year round, most conveniently on a cruise. Not to be missed: the famous Blue Hole, an indigo, coral-fringed gulf opening onto the turquoise blue of the Caribbean, and the gigantic gatherings of Nassau grouper in the spring.

### 4. FLORIDA
South of mainland USA, Florida stretches out its Keys archipelago, a group of islands and coral islets that has become a top-ranking American tourist location over the years. Wide stretches of the coast, from Key Largo in the north-west to Key West in the south-west, have been established as national parks to protect splendid coral gardens where large predators venture such as the tarpon, swordfish, marlin and shark, so dear to Hemingway. The winter, with its tropical storms, is to be avoided. Not to be missed: the John Pennekamp Marine Park, with its famous statue Christ of the Deep.

### 5. BAHAMAS
The Bahamas archipelago with its hundreds of islands, to the south of Florida, offers infinite diving possibilities. Naturally enough, the infrastructure is close to perfect: you can visit the Grand Bahamas to see the dolphins; Andros for its blue holes; New Providence for its shark shows; Harbor Island, Cat Island and Long Island for their wrecks, the quality of their coral reefs, the abundance of their fauna and, in particular, great concentrations of fishes. A real divers' paradise with many aspects which, thanks to the mildness of the Gulf Stream, can be visited all year round. Not to be missed: a meeting with the spotted dolphins of Maravilla.

### 6. DOMINICAN REPUBLIC
The Dominican Republic, which shares its island with Haiti, is the fashionable Caribbean location between Cuba and the Lesser Antilles. Although the Republic is enjoying a boom in tourism, diving is still not yet fully developed and thus very pleasant, especially in the north, around small islands bounded by a splendid coral reef, off the Samana peninsula. Easy diving except during the winter, when there is wind. Not to be missed: the freshwater cave at Cabrera, and the whales in Samana bay in the spring.

### 7. THE FRENCH ANTILLES
Diving in the French Antilles is well organised and here every year hundreds of divers at all levels discover the pleasures of the fauna and flora of the Caribbean. More enjoyable in the winter, the diving is mainly centred on Diamond Rock, or the wrecks of St Pierre bay (experienced divers preferred) in Martinique, around the Pigeon Islands and in the 'Cousteau Reserve' off Malendure, in Guadeloupe, and also at Saintes, Marie-Calante and St Martin. Not to be missed: le Sec Pate in the Saintes canal, the most beautiful dive in the area.

### 8. THE GRENADINES
The coral archipelago bordering the Lesser Antilles south towards South America is a vast territory where the diving is mostly in crystal-clear waters. Cruising from Bequia to Union Island passing by Canouan, Mustique or Mayreau, and it is a privilege to dive at the diminutive paradises of Palm Island, Petit Saint-Vincent and above all Tobago Cays, surrounded by well-preserved coral gardens. The best diving conditions are found on the west (Caribbean) coast at winter's end. Not to be missed: the unbeatable aquarium at Tobago Cays.

### 9. CONNEMARA
Protected from the cold north Atlantic by the terminal currents of the Gulf Stream, the coast of Connemara in Ireland, with its somewhat harsh surroundings, hosts good numbers of divers each year, many of them local. Beneath a vast canopy of laminaria weed is an astonishingly rich and varied fauna, in waters ranging from green to grey depending on the season and storms from the west. Diving takes place mainly at Inishbofin, Inishark, Inishturk, and Clare Island, preferably in September when the water is at its warmest and clearest. Not to be missed – nonetheless – the spring passage of pilgrim sharks, those giants of the sea, gathering to gorge on millions of salps flourishing in the early rays of the sun.

### 10. KERRY
In southern Ireland, Kerry cuts deep fjords into a coastline bordered with fields of algae waving in the current. At the confluence of the Atlantic and the North Sea, this region does not always provide the calm conditions with good visibility that one would hope for. Even at the height of the short, very popular summer season, local divers are used to wearing dry suits. This is the domain of enormous crabs and morays. Not to be missed: a face-to-face with the seals that swarm around some of the islands of the Skellings archipelago.

### 11. COSTA DEL SOL
In the south of Spain, the Costa del Sol benefits from the influence of rich Atlantic waters coming through the Straits of Gibraltar, permitting the vigorous development of a plentiful and varied fauna. The result is an astonishing mixture of Atlantic and Mediterranean fauna in water that is fairly cold, and not always very clear, except in the summer. Not to be missed: diving in the big blue sea in the middle of the Straits of Gibraltar, with the chance, however uncertain, of breathtaking meetings with dolphins, tuna, sharks, or whales.

### 12. COSTA BRAVA
In contrast with what happens on land, the coast of Spanish Catalonia to the north-west of the Iberian peninsula benefits from a surprising degree of protection from tourism, and all aspects of diving are flourishing. Numerous diving clubs allow you to go in search of the remarkable Mediterranean flora and fauna. Not to be missed: friendly encounters with the good-natured groupers of the Meras islands, which almost allow you to stroke them.

### 13. THE GULF OF LIONS
The French coast is not short of exceptional sites, whether on the Catalonian, Provençale, or Varois coasts. Strewn with a good number of marine reserves and an infrastructure of quality, this coast is the joy of French divers. It also attracts many

amateurs from northern Europe, delighted to swim during the summer months in the clear waters provided with richly decorated drop-offs and wrecks. It is preferable to wait until the end of the season to take the best advantage of superb glowing dives off Banyuls-sur-Mer, Marseilles or Saint-Raphael. Not to be missed: the fauna and flora of the national marine park at Port Cros, off Hyères.

## 14. CORSICA

The paradise of French diving is certainly to be found along the south and west coasts of the Isle de Beauté, somewhere between Calvi and Porto Vecchio. Warm limpid waters of an incomparable blue, clouds of tiny multicoloured fish, shoals of big silvery fish from out at sea, drop-offs populated with luminous gorgonias... Many professionals in love with their island will take you in search of the treasures of Scandola, Bonifacio straits or the Lavezzi islands. Not to be missed: Merouville, in the Lavezzi islands, where a semi-domesticated population of enormous good-natured groupers is in residence.

## 15. SARDINIA

To the south of Corsica, Sardinia prolongs her neighbour's ecological marvels in the blue of a Mediterranean ever more blue, ever more beautiful. If it has fewer reception facilities, the Sardinian coast is wilder and is an ideal place to look for precious red coral in all its splendour. Experienced divers travel to Sardinia from Italy, from September onwards, to profit from optimum water clarity before the storms of winter. Not to be missed: wonderful diving in the Corsica-Sardinia park off the north coast.

## 16. TUNISIA

Formerly famous for red gold, the famous Mediterranean coral, and for sponges, the coasts of Tunisia have an ancestral diving tradition. A tourist location *par excellence*, Tunisia welcomes new enthusiasts won over by warm, easy, colourful diving from late spring until the start of winter each year. In the north they prefer the environs of Cape Bon and Zembra island, and in the south the islands of Kerkenna and Djerba. Not to be missed: the caves of Tarbaka with their groupers and red coral.

## 17. SICILY

Undoubtedly Italy's most famous diving location, Sicily offers the thousands of islands that surround it, an adventure territory for the hundreds of divers, mainly Italian, who hurry there every year. Facilities abound and are of high quality: Sicily has plenty of colourful sites, caves and decorated cliffs for exploration in summer and autumn. Not to be missed: Scilla in the straits of Messina, with its carpets of red gorgonia and the creatures that rise at night from the depths to feed on plankton.

## 18. AEGEAN SEA

Diving was forbidden for a long time in Greece to protect its architectural heritage. The rare sites open to divers these days are not without charm. Sometimes they even allow an approach to fauna less nervous than elsewhere, such as loggerhead turtles, bream and octopus. The late season is naturally the best for this type of encounter, especially in the most popular islands in the Cyclades and Dodecanese. Not to be missed: the secret caves of the Sporades, giving shelter to the last monk seals in the Mediterranean.

## 19. ISRAEL

To the north of the Red Sea and the Gulf of Aqaba, around Eilat, live the most northerly corals in the world. A special world in miniature has developed along delicate reefs that now form part of a national marine park. Thanks to strict organisation, from Eilat one can visit Solomon's Garden, Moses' Rock and Coral Island, near the Egyptian border, all the year round. Not to be missed: an unforgettable dive with the semi-captive dolphins of Dolphin Reef.

## 20. EGYPT

The Egyptian Red Sea is indisputably one of the most beautiful diving areas in the world, first as regards attendance by European divers who find excellent facilities there, living as though on a cruise, at all levels and budgets. Some sites are restricted to experienced divers because of strong currents. There is no shortage of sites: around Sharm el Sheikh, the southernmost point of the Sinai (Tiran Strait, Ras um Sid, Ras Mohammed, Shark Reef, Ras Nasrani, Gordon Reef, Thomas Reef), d'Hurghada (Little Giftun, Sha'ab Abu, Umm Gammar, Umm Domm, Careless Reef), or off the south coast (Daedalus Reef, Brother island, Dolphin Reef, Fury Shoal), featuring splendid drop-offs and wonderfully colourful soft coral gardens. Not to be missed: the wreck of the *Thistlegorm* at the very tip of the Sinai.

## 21. SUDAN

Although reasonably well organised, Sudan is less visited and access is less reliable than in neighbouring Egypt. Underwater Sudan is mainly visited on cruises between October and mid-July. There are splendid preserved reefs to be discovered at Sha'ab Suedi, Sanganeb, at the foot of the famous lighthouse, Angarosh, Abingdon Reef and the wreck of the *Umbria*, richly decorated in glowing alcyonarians. Not to be missed: diving at Sha'ab Rumi to visit the remains of Commander Cousteau's underwater house (*Precontinent II*).

## 22. DJIBOUTI

At the confluence of the Red Sea and the Indian Ocean, the strait of Bab el Mandeb is entrance and exit to both marine territories with their manifold wealth. A small francophone area in the desolate countryside of the horn of Africa, Djibouti is certainly the place that serves as a vehicle for the imagination among Frenchmen keen on underwater adventure. For here indeed is adventure, in the waters of Goubet, Obock or the Gulf of Tadjurah, among the most thickly populated with fish in the world. The approach (big waves and strong currents, tricky sea) is sporting to say the least (to be avoided by beginners). Once underwater, what a sight! And at every dive, huge fish from the open sea... Not to be missed: the Seven Brothers archipelago with its giant loach.

## 23. TANZANIA

Off the coasts of East Africa, the Tanzanian islands of Pemba, Mafia and Zanzibar play between the Indian Ocean and the beginnings of the Mozambique channel. This is an unusual location, but is beginning to get organised. The mysterious drop-offs and multicoloured fish extend a welcome to you, on a stay or a cruise, between December and mid-April. Not to be missed: diving with the school of white dolphins at Pemba.

## 24. ALDABRA

A sanctuary for animal life in general, and marine life in particular, Aldabra is closely supervised and permission for access is grudgingly granted by the Seychelles authorities. A coral atoll situated between the Seychelles, to whom it belongs, and Madagascar, Aldabra can be visited by boat from Mahé, but the cost of the rare private or scientific cruises that might accept you is generally prohibitive. However, this true Grail is worth the sacrifice, such is the beauty of the seabed and the wealth of its incomparable fauna, especially when the waters are at their clearest in April and May. Not to be missed: diving after the sea turtle hatch, when the baby turtles invade the coastal waters in hundreds.

## 25. MADAGASCAR

This great Indian Ocean island has long been a famous diving location, especially in the north-west around the Nosy-Bé archipelago: Nosy-Komba and Nosy-Tanikely islands; towards the Mitsio archipelago: the island group made up of Four Brothers, Tsarabajina and the Ankarea bank; the Radamas archipelago with Russian Bay and the Iranja islands. The waters are clear from April to January: The reception facilities are perfect, and a rich and colourful fauna is to be seen. Not to be missed: the passage of humpback whales in the Sainte-Marie channel on the east coast.

## 26. SEYCHELLES

Off the coast of East Africa, this paradise of beaches for romantic lovers is also a prized diving location providing the comfort of numerous modern facilities for observing the colourful and varied fauna. The granite Seychelles (Mahé, Praslin, la Digue) are the territory of shoals of fish and manta rays; the coral Seychelles (Desroches, Alphonse) the domain of reefs, colourful drop-offs and the fauna of the open sea. You can dive there all year round, preferably in April–May and October–November. Not to be missed: gatherings of whale sharks in the autumn.

27. REUNION
Strangely little known, doubtless because coral is largely absent, in a maritime region where elsewhere it is normal. The diving at Reunion is worth more than just a detour. This French department in the southern Indian Ocean is much frequented by lovers of sporting adventure and gives access to an astonishing submarine domain. The facilities are excellent, and you can dive there all the year round, in search of dizzying drop-offs, caves, canyons richly decorated with coral and gorgonias, and often frequented by deep-sea fauna. Not to be missed: diving at Piton, swarming with a dense and colourful fauna.

28. MAURITIUS
The tourist location *par excellence* in the Indian Ocean, Mauritius has rich and colourful seabeds, and the means of visiting them at any season. As regards diving, there is a choice between the aquariums of the lagoon, with their multicoloured fauna and the drop-offs outside the reef where deep-sea fauna can be seen. Though the most visited region is Grand Bay and the Trou au Biches (the fabulous Whale Rock), you can try trips to gentle Trou d'Eau, Flic and Flac with its 'cathedral', Blue Bay, Belle Mare and its channel, or the shadow of the imposing silhouette of gloomy Brabant. Not to be missed: Plate island shark channel. Shivers guaranteed.

29. RODRIGUES
Less visited than its sister island Mauritius on which it depends, Rodrigues island is no less wild and beautiful. The coral blooms remarkably well, and harbours an abundant, colourful fauna that is more nervous and more authentic, both in the languid lagoon and on the outside of the reef. You can go there all the year round, flying from Mauritius. Not to be missed: the Cathedral off Port Maturin.

30. LAQUEDIVES
A wild archipelago off the west coast of India, Laquedives is not what you would call an organised diving location. Depending on the year, it is sometimes necessary to take everything you need with you, and careful planning is called for. The archipelago can be visited by boat from November to April (outside the monsoon season) and most of its coral reefs, with their abundant and extremely varied fauna, have never been dived. Not to be missed: the sharks swimming about in the channels.

31. MALDIVES
A diver's realm for more than thirty years. Perfect organisation and facilities, whether staying on the islands or cruising (the best idea), so one can set off 'blindfold' to explore this coral paradise south-west of India. A vast aquarium, the Maldives shelter a remarkable fauna, from clown fish to sharks and deep-sea manta rays. Its atoll make-up offers various diving possibilities in the lagoons, in channels, or on the external slopes of the reefs. You can dive there from November to April, avoiding the monsoon season. Not to be missed: dives around Ukulastila island where manta rays perform their unforgettable ballet.

32. SRI LANKA
This large island off the south-west coast of India is not particularly well known for diving. Even if it is not always easy to organise the diving, its coral-carpeted coasts and little-known reefs, such as those of Hikkaduwa on the west coast, hide some treasures not to be passed by. The monsoon season only permits submarine excursions during the winter. Not to be missed: gatherings of cachalot whales in the autumn on the west coast.

33. THAILAND
Sacrificed on the altar of mass tourism, some regions of the gulf of Thailand have suffered from excessive invasions, and the seabed bears the marks. Nonetheless there are still some beautiful sites to visit, basically on the west coast, off Phuket (Ko Phi Phi, Ko Rajah), in the Similan archipelago around the Surin islands, and the Burmese banks. The diving is well organised and takes place mostly in October because of the summer monsoons. One of the stars of the region is the harmless leopard shark, tame enough to allow itself to be approached. Not to be missed: gatherings of whale sharks at Richelieu Rock.

34. SONDE ISLANDS
The south Indonesian arc comprising the large islands of Sumatra, Java, Bali, Lombok, Sumbawa... is one of the world's richest biogeographical regions. In fact currents between the Pacific and Indian Oceans flow through the straits that separate these islands. The result is a wealth of remarkable reefs, of which Flores island is the finest example. More accessible and popular with tourists, Bali has numerous good quality diving sites. All the islands are reached by plane from Djakarta, and diving is mainly in the winter months, avoiding the monsoon season. Not to be missed: the drop-offs at Menjagan and the wreck of the USS *Liberty* off Bali.

35. SULAWESI
In a few years the northern part of this central Indonesian island pointing towards Borneo has become the Mecca of underwater photographers in search of marine microfauna. Thus, linked by plane with the Indonesian capital via Ujung Padang, the sea off Manado presents divers with a universe of colours and shapes of unusual wealth: the Bunaken marine park, the Sangihe archipelago (with its submarine volcanic activity), the Gangga islands and, above all, the strait of Lembeh, the most famous site. Dream-like dives practically all year round, with faultless organisation. Not to be missed – all the same – the Wakatobi site in the Tukang Besi islands at the extreme south-west of Sulawesi, one of the richest sites in the world.

36. KALIMANTAN
The Indonesian part of the great island of Borneo leads to a largely unexplored maritime area that promises adventure and discovery at almost any time of year. The best thing is to make a chain of flights from Djakarta to Berau on the island of Derawan on the north-east coast, and from there tour Sangalaki island with its manta rays, Kakaban with its prodigious drop-offs. Not to be missed: a dive in the jellyfish lake, in the heart of the Kakaban jungle.

37. SABAH
The Malaysian part of Borneo also gives access to an adventurous submarine territory of incredible richness, especially off the north coast. All year round, having reached Sandakan or Tawau via Kuala Lumpur and Kota Kinabalu by plane, then Semporna by car, you will find a welcome (after a few hours by boat) on the islands of Lankayan, Layang Layang and Sipadan. The reputation of this last has been earned by the quality of its dives (manta rays, turtles, sharks, not forgetting the famous ring of barracudas), as with the two tiny paradises, Pulau Mabul and Kapalai, reachable from it. Not to be missed: a dive in the immense underwater cave littered with sea turtle skeletons at Sipadan.

38. PALAWAN
One of the most beautiful islands of the Philippines separates the China Sea from the Celebes Sea to the south-west of the Philippine archipelago. Well-preserved fields of coral can be found there, particularly around the islands of Busuanga, Coron and Culion, not to mention the wrecks of ships sunk during the Second World War (experienced divers only). To be visited on a cruise, after flying from Manila, between October and June. Not to be missed: an unusual dive in the sulphurous lake of Cayangan, in turquoise waters at 40°C (104°F).

39. MINDANAO
The southernmost island of the Philippine archipelago, reached from Manila,

preferably between October and June, Mindanao borders the maritime territory of Cebu/Bohol, with fairly well-known sites such as Apo island, Panglao island or Cabilao island. To the south lie the marvels of the Celebes Sea and the Sulu Sea, across a scattering of heavenly islands (Basilan, Jolo, Tawitawi, Sitankai), difficult to reach and above all notorious for guerrillas and increasing piracy. To be avoided for this reason.

## 40. GREAT BARRIER REEF
The biggest reef in the world and the only living structure visible from the moon. A living monument, a giant natural marine park stretching 2,500 kilometres (1,550 miles) along the north-west coast of Australia. Naturally the wealth of fauna is remarkable, whether along the reefs (Ribbon Reef), around the islands (Heron, Lizard, Lady Elliot) or in the heart of the Coral Sea (Holmes Reef, Osprey Reef). Of course everything has been planned to bring this diving myth within your reach at any time of year. It can be reached from Townsville, Cairns or Port Douglas. Not to be missed: a visit to the enormous potato groupers at Cod Hole.

## 41. KANGAROO ISLAND
South of Australia, off Adelaide, Kangaroo Island is a unique flora and fauna reserve with all the attractions of Australia. Under the water, a mixture of South Pacific and southern Indian Ocean, with added icy subantarctic eddies, are unique living conditions for the most original, exuberant flora and fauna. Among the great laminaria, sea lions and camouflaged seahorses flee from the shadow of the predatory sharks that prowl in these emerald-glinting waters. Diving is best during the southern summer months, from October to April. Not to be missed: a search for the sea dragon looking like so many bits of seaweed.

## 42. PAPUA NEW GUINEA
Still largely unexplored, the reefs of Papua New Guinea yield new species to human curiosity. This is the realm of extravagantly coloured nudibranchs in particular. You can dive all the year round on the south coast off Port Moresby (Loloata, Eastern Fields), towards Milne Bay and Alotau, and on the north coast off Madang and the islands of New Britain (Walindi, Rabaul), New Ireland (Kavieng) or Entrecasteaux. A unique mixture in a world of exceptional fauna, with wrecks preserved from the last world war and active submarine volcanoes. A cruise is definitely not enough. Not to be missed: diving at Walindi, where 80 per cent of Indo-Pacific flora and fauna are 'listed'.

## 43. VANUATU
The former New Hebrides, north of Fiji, make up an assembly of atolls and sunken volcanoes that provide almost unlimited diving possibilities among reefs, and wrecks from the last world war, from Vanikoro to Tana, passing via Torres, Banks, Espiritu Santo, Eromango. Protected from excessive diving, the fauna of the South Pacific has developed well. Easily accessible all the year round although well away from the great tourist routes, Vanuatu is a remote paradise. Not to be missed: a visit to the *President Coolidge*, a steamship that sank in 1942; the largest wreck in the world accessible to divers.

## 44. NEW CALEDONIA
The biggest lagoon in the world extends its welcoming arms around this scrap of tropical France plonked off the coast of north-east Australia. In addition to the classic fauna of the Pacific, not to speak of the thousands of seashells in different colours and shapes, you can run into such unusual species as nautilus or dugong. There are many treasures off Noumea (the Amedée lighthouse, the wreck of the *Dieppoise*, the Boulari channel...) and off Port-Boisé (Prony, the Merlet reserve) to the south, Bouloparis and Koné to the west, Poum to the north, and Poindimié and Hienghene to the east. There are also islands with magic lagoons: Ouvea at Ile des Pins, passing through Lifou and Maré( An unforgettable memory that can be woven at any time of year. Not to be missed: diving right in the south at the Prony needle, which combines monumental coral with fauna from the open sea and the depths.

## 45. FIJI
As colourful as it is varied, Fiji's marine life is famous for its soft corals and multicoloured gorgonias, crystalline lagoons and giddy drop-offs. Easily accessible in any season, the Fiji islands have excellent facilities to help you discover the wonders of Beqa Island, Kadavu (Astrolabe Reef, Solo Reef) and Taveuni (Matagi and Qema Island). Not to be missed: diving in the Somosomo strait (Rainbow Reef and Great White Wall), and, at the foot of Taveuni, the biggest assembly of alcyonaria in the world.

## 46. TONGA
At the edge of one of the deepest ocean trenches on the planet, where submarine volcanoes spit, the Tonga Islands (Vava, Ha'apai, Tongatapu...) are mysterious from many aspects. The coral reefs have a wild appearance and the diving is not very well developed. An adventurous location that can be reached from Fiji all year round. Not to be missed: a face-to-face with the enormous potato groupers of Vava.

## 47. FRENCH POLYNESIA
In the middle of the South Pacific, offering well-organised reception and facilities, the various archipelagos of French Polynesia welcome you with open arms: Society Islands (Tahiti, Moorea, Bora Bora, Raiatea, Maupiti...), Gambier, Austales, Marquisas and Tuamotu, the divers' paradise with hundreds of atolls: Rangiroa, Fakarava, Manihi, Tikehau. Quite apart from the different sites (lagoons, channels, external reefs), it would be impossible to name all the species that swarm in this Neptune's Eden. You can go there at any time of year for unforgettable daily meetings with manta rays and sharks. Not to be missed: encounters with whales at Rurutu (Australes) at the end of the southern winter.

## 48. CALIFORNIA
Along the south-west coast of the USA diving is an institution. There is no lack of facilities or opportunities. And the field of adventure is not exactly the dullest in the world, because off the Californian coast grow submarine forests of giant kelp where you have the feeling of being an explorer on another planet. Here you can meet with an unusual fauna that ranges from seals to sea otters, blue sharks, sea angels, sea lions, grey whales, not to mention thousands of fish, including the famous Garibaldi orange, symbol of the region. The water is always cold, even during the best season, which is autumn. Not to be missed: the Monterey canyon, for the most beautiful submarine forests in the world.

## 49. GALAPAGOS
Protected by UNESCO, this enormous sanctuary for animal life in general and marine life in particular can be visited by boat from the coast of Ecuador, not without a degree of sentiment, in the wake of Darwin and his theory of the evolution of species. Thanks to powerful currents, some of the unique encounters that await the lucky diver include sea lions, penguins, marine iguanas, turtles, hammerhead sharks and manta rays. Despite the location below the equator, the island waters are relatively cold. Diving takes place all the year round, but preferably between October and December. Not to be missed: a prayer that this natural paradise remains so for ever.

## 50. COSTA RICA
Facing the Caribbean and the Pacific, this little Central American country is famous for a fabulous diving site off the Pacific coast around Coco Island, which can only be reached after thirty hours in a boat. But what a reward! The submarine fauna is incredibly dense, made up mostly by large species: hammerhead sharks, eagle rays, giant manta rays, whale sharks, dolphins... The quality of the site is as high as the cost of getting there — exceptional. You can dive there almost all year round, but because of the strong currents the site is only suitable for experienced divers. Not to be missed: enormous gatherings (up to three hundred at a time) of hammerhead sharks. We still don't know why.

Acknowledgements
The editor thanks all those who helped in the production of this work, particularly Patrick
de Wilde for his invaluable assistance and communicative zeal, and ARIEL FUCHS who
compiled the Diving Notebook.

CAPTIONS FOR THE OPENING PAGES
Pages 4–5: In the waters of the Bahamas, a mother dolphin protects and educates her small
offspring. They belong to the group of Atlantic spotted dolphins (*Stenella frontalis*),
considered by some scientists to be an entirely separate species, and by others a subspecies
of the tropical spotted dolphin (*Stenella attenuata*).
Pages 6–7: Well protected by the tentacles of its sea anemone (*Entacmaea quadricolor*),
the two-spined clown fish (*Premnas biaculeatus*) defends its hostess with determination.
The spines stand out from each cheek.
Pages 8–9: Here tropical life seems a kind of madness, with fruitful and unreal
invertebrates all packed together on a gorgonia branch: little coloured sponges, the ovoid
sacs of colonial sea squirts, and little hydra feathers…

The portraits of SOPHIE DE WILDE on page 13 are by PATRICK DE WILDE (left),
PASCAL KOBEH (centre) and JOELLE PICHON (right).

First published by Editions du Chêne, an imprint of Hachette-Livre
43 Quai de Grenelle, Paris 75905, Cedex 15, France
© 2002, Editions du Chêne – Hachette Livre
Under the title Le Bonheur sous la mer
All rights reserved

Language translation produced by Translate-A-Book, Oxford

© 2004 English Translation, Octopus Publishing Group Ltd, London
This edition published by Hachette Illustrated UK, Octopus Publishing Group,
2–4 Heron Quays, London, E14 4JP

Editorial direction   NATHALIE BAILLEUX
Artistic direction   SABINE HOUPLAIN
Layout   PATRICK DE WILDE
assisted by PASCAL BARAT
Corrections   ISABELLE MACÉ
Cartography   CYRILLE SUSS

Printed in Singapore by Tien Wah Press
ISBN: 1-84430-054-4